Dr. Jerzy Nowak
Department of Horticulture
Virginia Tech
301 Saunders Hall 0327
Blacksburg, VA 24061
E-mail: jenowak@vt.edu

Mutations, *In Vitro* and Molecular Techniques for Environmentally Sustainable Crop Improvement

Mutations, *In Vitro* and Molecular Techniques for Environmentally Sustainable Crop Improvement

Edited by

M. Maluszynski
Joint FAO/IAEA Division of Nuclear Techniques in Food and Agriculture, Vienna, Austria

and

K.J. Kasha
University of Guelph,
Department of Plant Agriculture,
Guelph, Canada

KLUWER ACADEMIC PUBLISHERS
DORDRECHT / BOSTON / LONDON

A C.I.P. Catalogue record for this book is available from the Library of Congress.

ISBN 1-4020-0602-0

Published by Kluwer Academic Publishers,
P.O. Box 17, 3300 AA Dordrecht, The Netherlands.

Sold and distributed in North, Central and South America
by Kluwer Academic Publishers,
101 Philip Drive, Norwell, MA 02061, U.S.A.

In all other countries, sold and distributed
by Kluwer Academic Publishers,
P.O. Box 322, 3300 AH Dordrecht, The Netherlands.

Selected papers have been published in *Euphytica*, Vol. 120, No. 3, 2001

Printed on acid-free paper

Printed in the Netherlands.

EDITORIAL

The views expressed in the papers, nomenclature used and way of presentation remain the responsibility of the authors concerned, who are also responsible for any reproduction of copyright material. In certain cases only a short communication is presented. All published papers were reviewed.

The excellent assistance of Mrs. Katayon Allaf and Kathleen Weindl, Joint FAO/IAEA Division, Plant Breeding and Genetics Section for editorial help, retyping some manuscripts and making corrections in English, is highly appreciated.

Contents

*Chapters indicated with an asterisk are reprinted from *Euphytica,* Volume 120, No. 3 (2001).

Preface

During the last thirty years, most increases in agricultural production were achieved through high input agriculture in areas with fertile soils and sufficient water. Intensive methods of production with high levels of nitrogen fertilizer and pesticides were often accompanied by environmental degradation and in some instances by pollution of the food supply. However, rapid population growth has also led to increasing use of marginal lands, where adverse soil and climatic conditions are serious constraints to food production. These areas are even more sensitive to ecological destabilization.

Environmentally sound systems of food production and land use are essential for meeting the food security needs of developing countries. To do this, greater genetic variability is needed within the best crop genotypes available for the areas in need coupled with better management practices and crop rotations. These requirements can only be realized if suitable crop varieties are bred. These should be varieties with a much shorter growing period, suitable for rotation, increased tolerance or resistance to diseases and pests as well as to drought and salinity and other adverse soil and climatic conditions.

Induced mutation techniques and other biotechnological approaches are major tools for creating variability for stress resistance or tolerance and for manipulating and evaluating them in efficient ways for improved germplasm. There are a number of them to choose from for producing new and useful variation within particular species and for particular agronomic situations. Additionally, scientists have become intensely interested in mutations as a means to widen and deepen our understanding of gene function. Among the many potential biotechnologies, *in vitro* culture and molecular methods seem most appropriate for developing countries..

The transfer of technology to developing countries needs more than just products of biotechnology but the technology itself. There are several reasons for this. First, the best target for improvement is local germplasm, which is already well adapted and has the specific traits acceptable to the consumers and thus can be selected by the local breeders. Second, it provides local employment and enhances local research capability, enabling breeders to also work on minor crops that otherwise would be ignored. However, this requires strong local government involvement and support which may not be optimal and hence the need for catalytic external assistance.

This publication contains the results of an FAO/IAEA Research Co-ordinated Project (CRP) entitled "Radiation Induced Mutations and Other Advanced Technologies for the Production of Crop Mutants Suitable for Environmentally Sustainable Agriculture". These are supplemented by information related to the topic on approaches to follow and methodologies available with emphasis on applications in developing countries. The publication deals only with seed crops, mostly cereals but also tropical legumes. Nevertheless the strategies used and conclusions reached are valid for many other types of crops.

Arising from the diversity of research objectives and strategies the outcomes of the project, although broad, are fundamentally twofold. First, a number of biotechnologies were developed and validated in association with the use of mutants to understand the genetics of specific plant traits; and second, a number of mutants were developed with beneficial characteristics and which have significant long term potential to contribute to sustainable agriculture. For example, within the first category, the use of advanced technologies to characterise specific mutated genes e.g. *ari-e.GP* responsible for semi-dwarfness plus salinity and drought tolerance in barley, and the development of applicable *in vitro* selection and haploid methods, means that the impact of the project will continue to yield benefits as the understanding of a particular gene function flows through to plant improvement programmes in many countries. Also, *in vitro* methods of doubled haploid production and the genetic information resulting from this project open new possibilities for large-scale production of radiation-induced doubled haploid mutants and high yielding doubled haploids from heterotic mutant crosses of barley and other cereals.

Within the second category, was the creation of radiation-induced mutants of rice, barley, wheat, tomato and buckwheat with increased tolerance to abiotic stress e.g. salinity, aluminium toxicity, cold tolerance, has direct relevance to crop improvement in Brazil, Bulgaria, Chile, China and Pakistan. Agriculture in many countries will also benefit from the development of disease resistant varieties as a result of technologies applied in this project. For example, the use of gamma radiation to generate asymmetric *in vitro* fusion lead to rice lines with resistance to blast disease, and the transfer of genes for resistance to wheat streak mosaic virus and barley yellow dwarf virus from wild relatives into crop cultivars.

James D. Dargie
Director, Joint FAO/IAEA Division
of Nuclear Techniques in Food and Agriculture
Vienna, Austria

Mutation genetics of salt tolerance in barley: an assessment of Golden Promise and other semi-dwarf mutants

B.P. Forster
Scottish Crop Research Institute, Invergowrie, Dundee DD2 5DA, Scotland, UK

Key words: barley, semi-dwarf mutation, *ari-e*.GP, salt tolerance

Summary

A review of research at the Scottish Crop Research Institute (SCRI) on the effects of semi-dwarfing genes on salt tolerance in barley is given. Work began in 1993 with the fortuitous and unexpected result that the cultivar 'Golden Promise' showed considerable tolerance to salt. Golden Promise is gamma-ray induced semi-dwarf mutant of the cultivar 'Maythorpe'. The parent and mutant cultivars are presumed to be isogenic, but show significant differences in their responses to salt stress. The positive and pleiotropic effects of the mutant gene, commonly known as *GPert* were found to be effective in a number of genetic backgrounds. Earlier, in 1991 Frackowiak showed that the *GPert* mutation was allelic to the *ari-e* mutants in barley. The *ari-e* mutants were salt tested and found to show the same positive responses to salt stress as Golden Promise. This supported the allelism tests, and consequently the *GPert* symbol was changed to *ari-e.GP*. The semi-dwarf mutant *sdw*.1 (also known as *denso*) and the *erectoides* semi-dwarf mutant, *ert-k*^{32} were also tested for their effects on tolerance to salt, but did not show any positive effects. Salt tolerance was therefore not a general phenomenon of semi-dwarf stature but specific to mutations at the *Ari-e* locus in these lines. Genetic markers (RAPDs, AFLPs and SSRs) have been used for fingerprinting, genetic mapping, and QTL analysis. The markers have helped: 1) confirm the isogenic relationship between Maythorpe and Golden Promise, 2) clarify the confusion over the pedigree of Golden Promise, and 3) genetically map the *ari-e.GP* locus and examine its pleiotropic effects.

M. Maluszynski and K.J. Kasha (eds.),
Mutations, In Vitro and Molecular Techniques for Environmentally Sustainable Crop Improvement, 1–16.

Background

Early work on salt tolerance at The Scottish Crop Research Institute (SCRI) was aimed at screening variation in cultivated (*Hordeum vulgare*) and wild (*H. spontaneum*) barley lines. There was a particular emphasis in screening lines which had been used in developing doubled haploid lines (DHs) as a segregating DH population could be exploited in genetic mapping of quantitative trait loci (QTLs) for salt tolerance. Various barley lines were subjected to a hydroponics experiment of two treatments, 0 and 175 mol m^{-3} NaCl, and salt tolerance measured in terms of shoot sodium content (mg Na/g dry weight, Forster *et al*., 1994). From this initial screen a breeding line HS26/55C was found to be the most salt tolerant spring barley. HS26/55C is an interesting line in that it was bred as a replacement for the popular, but mildew susceptible cultivar 'Golden Promise'. It was developed from a cross between cv. Golden Promise and a wild barley (*H. spontaneum* HS26) in an attempt to transfer mildew resistance from the wild species into Golden Promise (Thomas *et al*., 1988). It was therefore presumed (wrongly) that the salt tolerance of HS26/55C had been inherited from the wild line HS26 along with mildew resistance, as it was known that certain wild barley populations were well adapted to saline environments. HS26/55C was therefore tested along with its mildew resistant donor, HS26 and its recurrent parent, Golden Promise. After 4 weeks exposure to 175 mol m^{-3} NaCl the shoot sodium contents of HS26/55C and Golden Promise were found to be similar; 23.6 and 21.8 mg Na/g dry weight respectively (no significant difference). However, HS26 had more than twice the amount of shoot sodium (54.7, $P<0.001$, Forster *et al*., 1994). It was concluded that the salt tolerance of HS26/55C was inherited from Golden Promise, further work on Golden Promise was initiated.

The Golden Promise story

The *GPert* mutation was induced by gamma-ray irradiation of the cultivar 'Maythorpe' in 1956 (Fig. 1). The aim was to induce shorter stiffer straw in a good malting quality barley by mutation. The result was the production of the highly successful malting quality barley, Golden Promise, a direct mutant of Maythorpe (Miln Marster, described in Sigurbjornsson and Micke, 1969), *GPert* stood for Golden Promise *erectoides*.

Golden Promise offered barley growers a unique combination of highly desirable agronomic characters such as earliness, short stiff straw, easy combining ability, good resistance to grain and ear loss and to grain splitting with good malting quality. Golden Promise was particularly suited to agriculture of Northern Britain. In 1967 it was included on the recommended list of cereals for England, a year later it was recommended for Scotland where it dominated Scottish barley acreage from the 1970s to the mid 1980s. Golden Promise was susceptible to powdery mildew (*Erysiphe graminis* sp. *hordei*), but the widespread use of effective chemicals allowed Golden Promise to maintain its dominant position and take an increasing share of the UK market. The introduction of Golden Promise to Scotland revolutionised cereal growing to the benefit of the farmer, maltster, brewer and distiller (fungicide manufacturers also profited), and enabled Scotland to come to the fore as a producer of malting barley. Golden Promise became a standard for malt quality and is still grown today under contract for specialist distillers. The introduction of Golden Promise not only presented farmers with a new plant model (semi-dwarf stature), but also provided an additional morphological gene source for breeding. Golden Promise contains the recessive mutation *GPert* (Golden Promise *erectoides* – though not strictly an *erectoides* gene, see later: Influence of other mutant dwarfing genes).

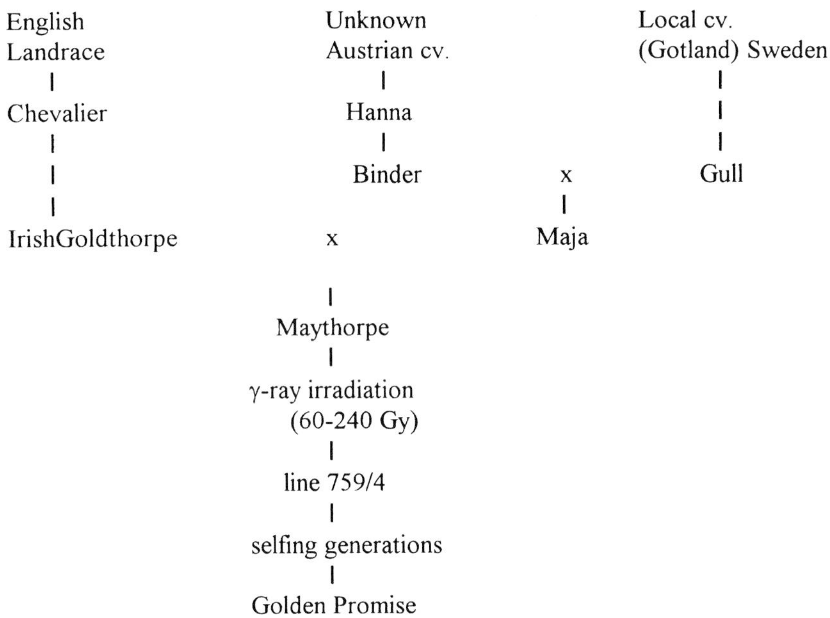

Figure 1. The pedigree of Golden Promise. Microsatellite analysis has resolved some of the confusion over the pedigree of Golden Promise and has confirmed that Irish Goldthorpe and not Goldthorpe (as stated in some pedigrees) is a grandparent (Russell *et al.*, 1997)

The mutated gene produces plants with short stiff straw and reduced awn length, it also has pleiotropic effects on yield and grain size. Golden Promise was an extremely successful cultivar and the *GPert* mutation has been bred into a number of successful cultivars e.g. 'Midas' (1970), 'Goldmarker' (1975), 'Goldspear' (1976), 'Celt', 'Chad' and 'Tyne' (1987). Also 'Goldfield' (1969), 'Minak' (1976), 'Yamina' (1979), 'Ingot' (1980), 'Doublet' (1983), 'Esk' (1985), 'Everest' (1985), ''Fleet' (1985), and 'Ayr' (1986) were placed on the UK National List. Golden Promise has been one of the most important induced mutants in the history of agriculture. The question was why should Golden Promise, a cultivar adapted to northern parts of the UK exhibit tolerance to salt?

Salinity worldwide

Soil salinity is one of the most serious stresses in agriculture. It has been estimated that about one billion hectares of the world's land is affected by salt, sixty percent of which is cultivated (Epstein *et al.*, 1980). Saline soils occur naturally along coastlines and in arid and semi-arid regions. They are most widespread in areas that combine low rainfall with high surface evaporation and poor drainage. Secondary salinity can be caused by irrigation with brackish water especially if there is a high water table. The problem is an ancient one: salinity build up in irrigated fields and drains caused the collapse of Mesopotamian agriculture and civilisations such as the Sumerians some 5000 years ago (Pearce, 1987; Jacobsen and Adams, 1968). The lesson of Mesopotamia has not been learnt. Today it is estimated (FAO) that half of the world's farms have been damaged by salt (Pearce, 1987). Irrigation water can deposit 1-10 tonnes of salt per hectare per year (SCOPE, 1978). In Pakistan, 2 tonnes of salt are added to every irrigated hectare every year (Stoner, 1988). In the relatively short history of intensive cropping in Australia salinisation caused by irrigation has taken 2×10^5 hectares out of production, for N. America the figure is 3.2×10^4 (SCOPE, 1978).

Large-scale irrigation/drainage schemes such as the Pakistani Left Bank Outfall Drain along the river Indus can help alleviate soil salinity, but these civil engineering schemes are huge and are prohibitively costly for many developing countries. They can also cause problems further downstream. In 1970 the American Bureau of Reclamation spent over $300 million building a chemical plant by the Colorado River to remove salt which had drained in from local fields. Previously, the salt deposited in the Colorado flowed south into Mexico where it caused soil salinity problems and crop failure. The plant was built to appease Mexican farmers and aimed to recycle water and remove saline water via long drainage canals, which connected to the sea (Pearce 1987).

An alternative is to improve the salt tolerance of crop plants. Salinity affects plants by reducing growth. The physiology and biochemistry of salt tolerance is poorly understood, but it is known that some crop species are better suited to saline environments than others; sugarbeet (*Beta maritima*), for example, is derived from a halophytic progenitor. In other crop plants considerable genetic variation exists which can be exploited to improve salt tolerance. Adapting crop plants to saline environments is a major challenge, and the identification of genes controlling responses to salt stress are of great interest.

Verification of the isogenic relationship between Maythorpe and Golden Promise

Although seed of Golden Promise is still widely available, Maythorpe its parental line has not been grown commercially since the 1950s, and seed was not available from Zeneca Seeds who took over Miln Marsters in the mid 1980s. Seed of Maythorpe was obtained from Dr M. Ambrose curator of the UK germplasm collection, John Innes Centre, Norwich. It was necessary to confirm the authenticity of extant Maythorpe seed and therefore DNA profiling was carried out. Comparisons of genetic fingerprints of Golden Promise and Maythorpe were compared using randomly amplified polymorphic DNAs (RAPDs, Figure 2). Analyses of more than 300 RAPD profiles (1200 strong bands) showed no variation between the two cultivars, the DNA profiles being identical in all cases (Pakniyat, 1996). The result was later verified using amplified fragment length polymorphisms (AFLPs, with one exception, see section on AFLP variation between Golden Promise and Maythorpe) and simple sequence repeat fingerprinting (SSRs, Russell *et al.*, 1997). The lines were apparently isogenic, and any differences detected between them may be attributable to the gamma-ray treatment, with obvious implication for *GPert*, as this is the only known mutation in Golden Promise.

Figure 2. Comparisons of paired RAPD fingerprint tracks of Golden Promise and Maythorpe (from Pakniyat, 1996). Primers used, from top left to bottom right, K9, K8, K6, K1, S2, P11, P9, P3, N13, M9, F11, E16, E5, S10, K7 and K3. S2 and S10 are SCRI primers, the rest are Operon primers

Comparisons of Golden Promise and Maythorpe for shoot Na^+ content and yield in salt stress

In the Triticeae, which includes the small grain cereals, there is a general association between salt tolerance and low shoot sodium content. Previous experiments (Forster *et al.*, 1994) showed that Golden Promise was able to limit the amount of sodium (Na^+) taken up during salt stress compared to its isogenic parent Maythorpe. The early experiment was conducted on young plants, it was therefore important to see whether shoot Na^+ content of vegetative shoots was related to yield. Plants of Golden Promise and Maythorpe were grown in control (25 mol m^{-3} NaCl) and salt stress (150 mol m^{-3} NaCl) as described by Forster *et al.* (1994) using a split plot design. After 4 weeks at treatment concentrations half the plants were harvested for shoot Na^+ analysis. The rest were grown to maturity and measurements taken on: height, main stem weight, mean number of spikes/plant, mean number of seeds/spike, mean number of sterile spikelets/spike and seed weight/plant (Table 1).

Table 1. Comparison of Golden Promise and Maythorpe for shoot Na^+, grain yield and other characters (From Pakniyat, 1996)

Genotype	NaCl (mol m^{-3})	Height (cm)	Main stem weight (g)	Number of spikes per plant	Number of seeds per spike	Number sterile spikelets per spike	Seed wt/plant (g)	Na^+ mg/g of dried shoot
Golden Promise	25	84.3^c	2.3^b	12.1^b	17.1^c	6.1^a	7.7^c	6.2^a
Golden Promise	150	55.5^a	0.9^a	4.6^a	11.5^b	9.1^b	3.8^b	16.8^c
Maythorpe	25	113.5^d	3.4^c	12.3^b	20.8^d	6.2^a	11.8^d	8.2^b
Maythorpe	150	64.0^b	1.1^a	3.3^a	6.7^a	12.0^c	2.4^a	21.4^d

Column means followed by the same letter are not significantly different ($P<0.01$).

From Table 1 it can be seen that Golden Promise and Maythorpe were significantly different in height for both treatments, which is not surprising as this is a major effect of the dwarfing gene in Golden Promise. The number of seed/spike was significantly higher for Maythorpe in the control, but in salt stress this was reversed. Similarly, seed weight/plant was higher in Maythorpe than Golden Promise in the control, but in salt stress Golden Promise produced higher grain yield/plant.

Golden Promise also had significantly less sterile spikelets/spike in salt stress treatment than Maythorpe. Salt stress also reduced the number of florets per ear. As in previous work shoot Na^+ content of Golden Promise was significantly less than Maythorpe, and this was found to be significantly correlated with seed weight/plant ($r = 0.86$). Salt stress had significant effects on shoot Na^+ number of seeds/spike, height, number of sterile spikelets/spike and seed weight/plant. By all criteria used Golden Promise performed better in salt stress compared to Maythorpe. The most beneficial effect was the reduction in the number of sterile spikelets in Golden Promise compared to Maythorpe.

Comparison of cation discrimination between Golden Promise and Maythorpe

Golden Promise has been shown to be more salt tolerant than Maythorpe in terms of yield and shoot Na^+. An experiment was set up to see if the ion exclusion in salt was Na^+ specific. A hydroponics experiment was therefore set up in which Golden Promise and Maythorpe were grown in control and salt stress, 0 and 150 mol m^{-3} NaCl respectively. After 4 weeks of each treatment shoot cation (Na^+, K^+, Ca^{2+}, Mg^{2+} and P^{3+}) concentration was measured using an ACP mass spectrometer (Table 2).

Table 2. Cation concentrations in shoots of Golden Promise and Maythorpe in control (0) and salt (150 mol m^{-3} NaCl) treatments (Data extracted from Pakniyat, 1996)

Genotype	NaCl (mol m^{-3})	Ion concentration (mg/g) Na^+	K^+	Ca^{2+}	Mg^{2+}	P^{3+}
Golden Promise	0	0.2^a	25.3^a	1.2^a	0.7^a	5.1^a
	150	9.1^b	16.6^b	1.4^b	0.3^b	4.2^b
Maythorpe	0	0.2^a	24.1^a	1.3^a	0.8^a	6.1^a
	150	13.0^c	17.6^b	1.5^b	0.4^b	4.3^b

Means in each column followed by the same superscript are not significantly different.

As in other experiments, Golden Promise again showed significantly less shoot Na^+ than Maythorpe in salt stress conditions. However, the shoot concentrations of K^+, Ca^{2+}, Mg^{2+} and P^{3+} showed no significant differences between the two cultivars for either treatment (Pakniyat, 1996). The only difference between Golden Promise and Maythorpe was for shoot Na^+ in the salt treatment and indicates that the cation exclusion effect of Golden Promise is specific to Na^+ in saline conditions.

Comparison of responses of *GPert* and non-*GPert* barley lines to salt stress

Since the release of Golden Promise in 1966 the *GPert* mutation has been incorporated into many new cultivars and breeding lines of barley. It was therefore interesting to see whether the pleiotropic effect of *GPert* in limiting shoot Na^+ uptake was expressed in these different and more modern genetic backgrounds. This was assessed by comparing the shoot Na^+ contents in spring barley lines carrying *GPert* with their non-*GPert* parents. Twenty-one *GPert* and 20 related non-*GPert* lines were tested for shoot Na^+ content after 4 weeks of exposure to high (175 mol m^{-3} NaCl) and low (25 mol m^{-3} NaCl) salinities. The results have been formally published by Pakniyat *et al.* (1997b) where pedigrees can be found. There were few differences between the lines at the low salt treatment, but at the high (175 mol m^{-3} NaCl) shoot Na^+ range from 'Clansman' with 14.6 to 'Claret' with 30.8 mg Na^+ g^{-1} dry shoot. The Golden Promise/Maythorpe comparison was as standard in that the shoot Na^+ content of Golden Promise was significantly less than that of Maythorpe in salt stress conditions. Also, Midas, Clansman, Ayr, 'Forth', 'Ingot' and 'Zarina' (*GPert* lines) contained significantly less shoot Na^+ ($p<0.01$) than their non-*GPert* parental lines; 'Proctor', 'Grand Prix 4', 'Athos', 'Akka', 'Lofa Abed' and Maythorpe respectively in salt stress conditions. The nine top performing lines were all *GPert* types and the eight worst lines all lacked the *GPert* mutation. Group comparisons of the *GPert* versus the non-*GPert* lines showed significant differences, the mean shoot Na^+ contents of the *GPert* carrying group being significantly less than those of the non-*GPert* group for both treatments ($p<0.01$). A highly significant difference was found between the two groups with the best performing lines carrying the *GPert* mutation and the worst lacking it. The results confirmed a positive effect of *GPert* on salt tolerance in reducing shoot Na^+ and that this was effective in a range of spring barleys, but that this was also regulated by genetic background (Pakniyat *et al.*, 1997b).

Influence of other mutant dwarfing genes

Mutation has been defined as a change in the heritable constitution, and is an important source of genetic variation in living organisms. Along with recombination, mutation is one of the major driving forces in evolution. Mutation occurs naturally at very low frequencies, in barley as in many other species mutations can be induced at much higher frequencies using physical (Amer and El-Rassas, 1994), chemical (Kapala and Rybinski, 1995) and biological mutagens (Kumar, 1996). Barley has a long record as being an experimental plant for mutation research. Stadler (1930) used barley as a model to demonstrate accelerated mutation rates in his work with ionising radiation. In the mid-1930s barley was intensively used in breeding experiments notably at Svalöv in Sweden, after the pioneering work of Müller (1927) and Stadler (1930). In Sweden a widespread research programme was set up to develop useful barley mutants; by 1986 some 9000 had been characterised (Lundqvist, 1986). Most of these are maintained in gene banks such as the Nordic Gene Bank, Alnarp, Sweden, and are listed in the GrainGenes database.

With respect to breeding, it was hoped that mutations induced in the best local cultivars would quickly produced improved lines. However, the detection of mutants among offspring of treated seed proved to be laborious. Many induced mutations have direct or indirect adverse effects. In addition, many desirable mutations are recessive and hence are difficult to track in a breeding programme. According to Briggs (1978) less than 1% of all observed mutations are of value to breeders. Nevertheless, some very important mutations have been utilised in breeding, *GPert* being a prime example.

The *GPert* symbol is a misnomer as Golden Promise is not a true *erectoides* type. *Erectoides* types have reduced internodes, but there is huge variation in the observed heights, our observation on Golden Promise is that it has a reduced rachis internode length compared to Maythorpe (Thomas, unpublished data). The *GPert* mutation is recessive and has been mapped (Thomas *et al.*, 1984) to barley chromosome 7 (5H) near to the centromere. The *Ari-e* locus located on chromosome 7 (5H) (Persson, 1969), has four well-defined mutants: *ari-e*.1, *ari-e*.119, *ari-e*.156 and *ari-e*.228. These mutants were produced by various treatments in the cultivars Foma and Bonus (Table 3); they are named *breviaristatum* (short awned) and have similar ear morphologies to Golden Promise (Figs 3 and 4).

Table 3. Semi-dwarf mutant genotypes used, their parental lines, mutagens and reference

Mutant line	Parent	Mutagen	Reference
ari-e.1	Bonus	Neutron (180 rad)	Kucera *et al.* 1975
ari-e.119	Foma	X-ray (7200 rad)	Kucera *et al.* 1975
ari-e.156	Foma	Propylmethanesulfonate	Kucera *et al.* 1975
ari-e.228	Foma	N-methyl-N-nitrosourea	Kucera *et al.* 1975
Golden Promise	Maythorpe	gamma-ray (6-24 rad)	Sigurbjornsson and Micke, 1969
Diamant	Valticky	X-ray	Bourma, 1967
Alf	Bomi	Neutron	Haahr and von Wettstein 1976
Pallas	Bonus	X-ray	Borg 1959

Reproduced by kind permission of Euphytica, Pakniyat *et al.* 1997a.

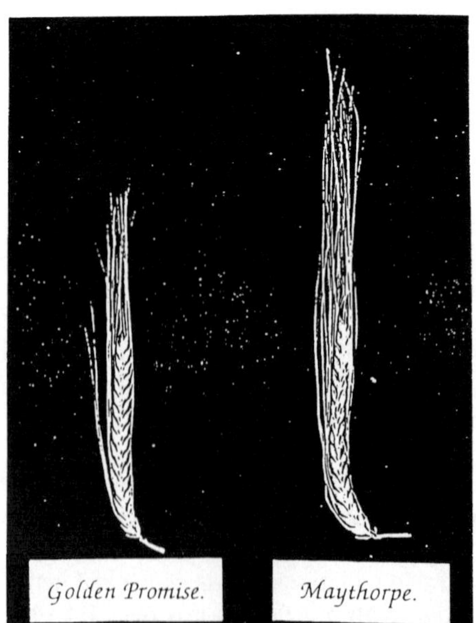

Figure 3. Comparison of ear morphology of Golden Promise and Maythorpe

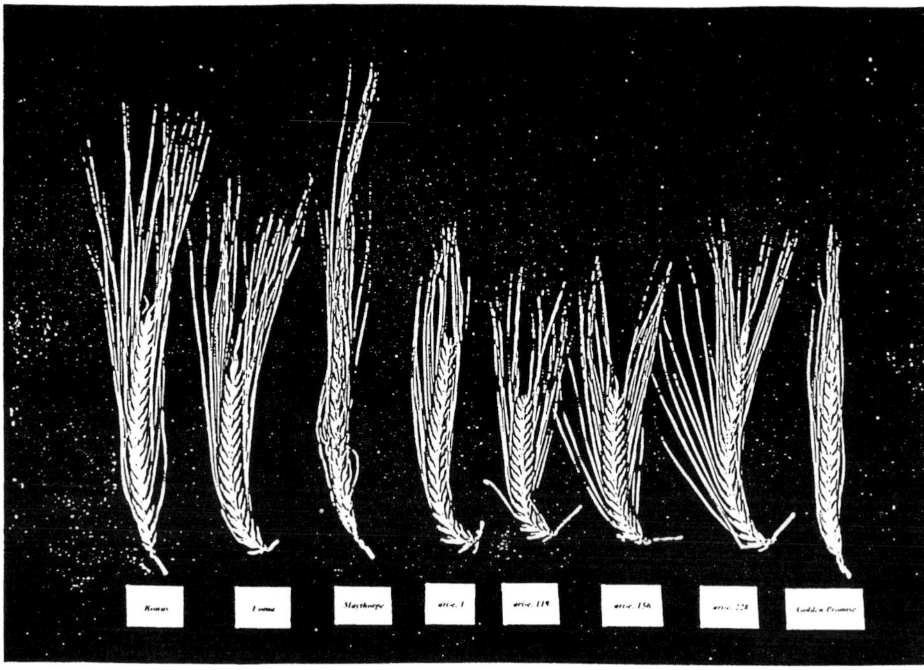

Figure 4. Comparison of ear morphology of Golden Promise, *ari-e* mutants and respective parental lines

Franckowiak (1991) demonstrated that the *ari-e* mutants are allelic to *GPert*. An experiment was therefore set up to see if the *ari-e* mutants responded in the same manner to salt stress as *GPert*. These mutants therefore were tested along with their isogenic non-mutant parents in high (175 mol m^{-3} NaCl) and low (25 mol m^{-3}) salt hydroponics. In addition, two other semi-dwarf mutant types were tested; 'Pallas' a direct mutant of 'Bonus' which carries the *erectoides* mutant, *ert-k*^{32}, and 'Diamant' and 'Alf' which carry the *sdw*.1 (*denso*) mutation. Diamant and Alf are also direct mutants, Diamant of 'Valticky' and Alf of 'Bomi'. The *ert-k*^{32} locus has been mapped to chromosome 6 (6H, Gustafsson *et al*., 1971), *sdw*.1 has been mapped to chromosome 3 (3H, Barua *et al*., 1993; Laurie *et al*., 1993), they are therefore independent of the *ari-e* and *GPert* mutations on chromosome 7 (5H). Salt tolerance was measured in terms of dry matter production, shoot Na^+ content and shoot carbon isotope discrimination ($\delta^{13}C$). Carbon isotope discrimination was used as an index of stress response, the more stressed a plant has been the less negative the $\delta^{13}C$ value (Farquhar and Richards, 1984).

The results of semi-dwarf mutant responses to salt stress are described in detail by Pakniyat *et al*. (1997a); a summary is given here. The four *ari-e* mutants and Golden Promise showed significantly better tolerance to salt in terms of shoot Na^+ and $\delta^{13}C$ values and dry matter production compared to the other mutants and their respective non-mutant parents. The *sdw* mutants Diamant and Alf, and *ert.k*^{32} mutant Pallas did not perform differently from their respective parental lines, i.e. these mutants were inert with respect to salt tolerance. The $\delta^{13}C$ values of *ari-e* mutants and Golden Promise were significantly less negative than *sdw*.1 (Diamant and Alf) and *ert-k*^{32} (Pallas) mutants in salt stress conditions (Fig. 5). The results showed that semi-dwarfism is not generally associated with salt tolerance, but specific to *ari-e* and *GPert* mutants. This may be associated with different modes of action of the various semi-dwarfing genes. The *ari-e* and *GPert* mutations produce semi-dwarf phenotypes with short awns. The *sdw* mutant is characterised by semi-prostrate growth and shortened basal internodes, there is no effect on awn length. The *ert-k*^{32} mutation is a true *erectoides* type, characterised by reduced internode length, dense spike and erect habit. Since the *ari-e* and *GPert* mutations gave the same responses to salt stress the results supported Franckowiak's finding (1991) that these are allelic. The *GPert* symbol was therefore changed to *ari-e.GP* (Pakniyat *et al.*, 1996); the new symbol will be used henceforth.

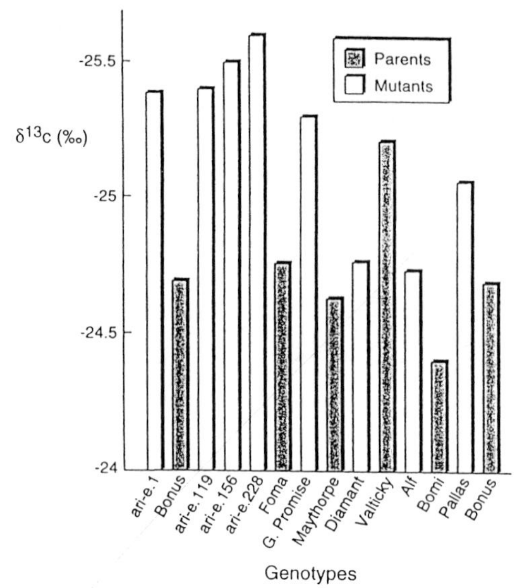

Figure 5. $\delta^{13}C$ values of barley mutants and their parents after salt stress (175 mol m^{-3} NaCl)

Response of semi-dwarf mutants to gibberellic acid

Gibberellins are a class of plant hormones, which in cereals can have a pronounced effect on stimulating shoot elongation. Gibberellic acid (GA_3) application to barley can produce an observable effect in tall genotypes, but a much greater effect can be found in dwarf mutants, which have low endogenous levels of gibberellins (Zwar and Chandler, 1995). There are, however, both GA_3-sensitive and GA_3-insensitive dwarf types of barley (Hanson *et al.*, 1980; Boulger *et al.*, 1981; Francowiak, 1991). An experiment was therefore set up to investigate seedling reactions of the various induced semi-dwarf mutants to exogenous GA_3, and to see whether the reactions grouped in a similar manner to that of their salt stress responses. The work has been published in Pakniyat *et al.* (1997a) and summarised below.

Seedlings were grown in hydroponics solution containing 50 ppm GA_3. After 10 days shoot length was measured and compared to shoot length and coleoptile length in non-GA_3 conditions, the shoot length ratios of GA_3+/GA_3- are given in Table 4.

All the *ari.e* mutants (including Golden Promise) showed very little response to exogenously applied GA_3, their shoot and coleoptile ratios were close to unity (Table 4). The parental lines of the *ari-e* mutants, Bonus, Foma and Maythorpe, were sensitive, as were the *ert-k*^{32} mutant (Pallas), the two *sdw.*1 mutants (Diamant and Alf) and their respective parental lines Bonus, Valticky and Bomi respectively. The data confirm previous results (Hanson *et al.*, 1980) that Golden Promise and other *ari-e* lines are relatively insensitive to exogenous gibberellic acid. Gibberellin insensitivity is probably involved in the phenotypic expression of semi-dwarf stature and short awns of *ari-e* carrying genotypes; semi-dwarfism caused by *sdw.*1 and *ert* is probably effected by another mechanism. It is not known whether gibberellin insensitivity plays a direct role in salt tolerance of *ari-e* lines, if this was established a seedling GA_3 test could be developed as a rapid and cheap screen for salt tolerance. There is, however, a vast array of gibberellins that may affect responses to salt stress.

Table 4. Seedling shoot length ratio GA_3+/GA_3- of mutant and parental lines

Mutant	GA_3+/GA_3- ratio		Parent	GA_3+/GA_3- ratio	
	Shoot	Coleoptile		Shoot	Coleoptile
ari-e.1	1.09	1.01	Bonus	1.39	1.12
ari-e.119	1.09	1.01	Foma	1.35	1.11
ari-e.156	1.09	1.04	Foma	1.35	1.11
ari-e.228	1.09	1.02	Foma	1.35	1.11
Golden Promise	1.08	1.02	Maythorpe	1.48	1.15
Pallas	1.33	1.16	Bonus	1.39	1.12
Diamant	1.45	1.13	Valticky	1.52	1.12
Alf	1.22	1.14	Bomi	1.27	1.16

Reproduced by kind permission of Euphytica, Pakniyat *et al.* (1997a)

AFLP variation between Golden Promise and Maythorpe

It was stated earlier that genetic fingerprinting using RAPDs and SSRs showed no differences between Golden Promise and Maythorpe. However, during routine genome screening of barley genotypes with AFLPs a single band shift was detected between Golden Promise and Maythorpe (Fig. 6). AFLP bands were generated for Golden Promise and Maythorpe using template DNA that had been derived from DNA digested with the restriction enzymes *Pst*I and *Mse*I. Standard AFLP methodology was used (Zabeau and Vos, 1993; Vos *et al.*, 1995). Seven *Pst*I/*Mse*I primer combinations were used to screen the two cultivars, giving a total of 490 bands of which one (from P23M81) was polymorphic and showed a one base-pair difference between the two lines (Fig. 6). This gave a polymorphism level of 0.2%, typically a level of 11-22% can be expected between barley cultivars.

The AFLP variation between Golden Promise and Maythorpe is a recent finding and requires verification. It has yet to be cloned and mapped. If the band shift is genuine and is associated with the *ari-e.GP* mutation, the AFLP would be expected to map to the *ari-e* locus on chromosome 7 (5H). In addition other *ari-e* mutants would also be expected to show variation for this AFLP band. If confirmed the AFLP would be extremely useful as it could be used to sequence the mutant gene. The observed AFLP variation could however be another phenotypically undetected, mutational difference between the two lines.

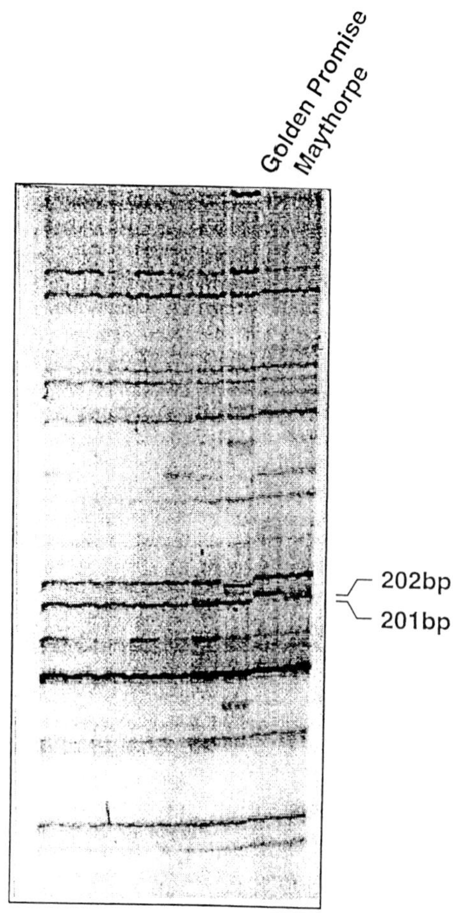

Figure 6. AFLP variation between Golden Promise and Maythorpe shows a band shift of one base pair

Quantitative trait loci (QTL) associations of *ari.eGP* and *sdw* with respect to salt tolerance

At SCRI a barley doubled haploid (DH) population has been produced which segregates for both *ari-e.GP* and *sdw*.1. The population consists of 156 DHs derived from the cross Derkado x B83-12/21/5, Derkado carries the *sdw* mutation, B83-12/21/5 carries *ari-e.GP*. This DH population provides interesting material in which to study the contrasting effects of these commercially important dwarfing genes in combination (double dwarfs), separately (*ari-e.GP* versus *sdw*.1), and in absence (tall genotypes) in similar and modern genetic backgrounds.

DNA was isolated from DH seedlings and analysed using sequence-specific amplified polymorphisms (S-SAPs, AFLPs and SSRs). Linkage maps were constructed from 240+ markers with the Joinmap 2.0 ™ programme, and include precise map locations of *ari-e.*GP and *sdw.*1. The DHs were simultaneously tested in control (0 mol m^{-3} NaCl) and salt stress (150 mol m^{-3} NaCl) hydroponics. Physiological responses to salt stress were treated as quantitative traits and analysed for associations with mapped genetic markers. For QTL mapping the number of markers was restricted to 120+, spaced at least 1 cM apart.

Data analyses of these experiments are not yet complete, and will be prepared for publication elsewhere. However, some preliminary results can be reported. To date, the most significant result is that for shoot $\delta^{13}C$. This was affected by seven QTLs with the most notable effects coming from those that map to the *sdw.*1 and *ari-e.GP* loci. The effect of the *sdw.*1 locus is to make the $\delta^{13}C$ value more negative indicating greater susceptibility to salt. This was also true for the Derkado tall, wild type allele at the *Ari-e* locus indicating that the mutant allele, *ari-e.GP* confers significantly greater tolerance to salt. Shoot $\delta^{13}C$ showed more heritability than other characters analysed to date (tiller number, growth stage, shoot weight, nitrogen content, $\delta^{15}N$ and carbon content – shoot Na^+ and Cl^- contents are not yet analysed). The implication is that a large portion of the phenotypic variation for salt tolerance in the material studied is genetic and can be manipulated by selection for the *ari-e.GP* dwarfing gene. It also follows that selection for the tall wild type allele at the *Sdw* locus would also be beneficial. However initial data show a positive interaction between the two mutant dwarfing genes with double dwarf genotypes (*ari-e.*GP/*sdw.*1) performing well under salt stress.

Drought tolerance and other stresses

In our experiments on salt tolerance we have used carbon isotope discrimination ($\delta^{13}C$) as a measure of stress response. This has been measured in both stress and control conditions. The *ari-e* mutants always show more negative $\delta^{13}C$ values than their non-mutant isogenic parents do, in all treatments. This indicates that the genetic control of carbon isotope discrimination in *ari-e* lines is constitutive and that these lines are more water use efficient. The implication is that *ari-e* mutant lines are relatively tolerant to drought as well as salt. This would not be surprising, as salt and drought tolerances are often associated. It would therefore be appropriate to test *ari-e* mutants for responses to a range of environmental stresses.

Golden Promise in biotechnology

Transformation of barley is currently in a developmental phase with various procedures and genotypes being explored. Golden Promise features prominently in this early work as large numbers of transgenics can be produced (Wan and Lemaux, 1994). Golden Promise has thus become a standard genotype for barley transformation with immature embryos, callus and microspores all being used as targets for plasmid bombardment. One explanation for success with Golden Promise is that it is a responsive genotype in tissue culture; cultures from various sources can be transformed and then manipulated to regenerate high numbers of green fertile plants. It is unclear whether transformation/tissue culture success of Golden Promise is associated with the *ari-e.GP* mutation though a simple comparison with Maythorpe and other *ari-e* isogenics would help determine this. It is clear, however, that Golden Promise is a useful cultivar for barley transformation and can be used as a starting point for breeding transgenes into other better adapted and more contemporary local types. Doubled haploidy via anther culture and microspore culture will be a key process in accelerating the production of desired homozygous recombinants for breeding purposes, and again the responsiveness of Golden Promise to anther/microspore culture will facilitate this. Genetic knowledge of Golden Promise can also be exploited. Because Golden Promise has been an extremely important cultivar it has been the subject of many genetic studies including pedigree analysis and genome scanning, its genotype is well defined. Golden Promise specific genetic markers can be used as selection tools (to select for, or to select against Golden Promise genes and genomic regions) in marker assisted breeding. Hopefully, Golden Promise will continue to live up to its name.

Acknowledgements

The work described has been sponsored in part by contract No. 7860/R2 with IAEA/FAO and carried out in collaboration with the Cell and Molecular Genetics, Cellular and Environmental Physiology and Chemistry departments at the Scottish Crop Research Institute. It has involved Dr W. Powell, Dr R.P. Ellis, Dr W.T.B. Thomas, Dr H. Pakniyat, Dr C. Simpson, and Dr L.L. Handley, with technical help from E. Baird, C. Scrimgeour, R. Keith, D. Gordon and S. Holdus. The SCRI receives grant-in-aid from the Scottish Office Agriculture, Environment and Fisheries Department.

References

Amer, I.M. and El-Rassas, H.N. (1994). Utilization of gamma irradiation in barley breeding. Annals of Agricultural Science (Cairo) **39**: 257-268.

Barua, U.M., Chalmers, K.J., Thomas, W.T.B., Hackett, C.A., Lea, V., Jack, P., Forster, B.P., Waugh, R. and Powell, W. (1993). Molecular mapping of genes determining height, time to heading and growth habit in barley (*Hordeum vulgare*). Genome **36**: 1080-1087.

Borg, G. (1959). Svalöf's original Pallas barley (Sv.04032), a new 2-row spring barley produced by X-ray treatment of Svalöf's Bonus barley. Sveriges Utsadesforenings. Tidskrift. 72-96.

Boulger, M.C., Sears, R.G. and Kronsrad, W.E. (1981). An investigation of the association between dwarfing sources and gibberellic acid response in barley. In: Barley Genetics IV. Proceedings of the Fourth International Barley Genetics Symposium, Edinburgh, pp. 550-553.

Bouma, J. (1967). New variety of spring barley, 'Diamant' in Czechoslovakia. In: Induzierte Mutationen und ihre Nutzung, Erwin Baur Gedachtnisvorlesungen IV, 1966, Abh Dt Akad Wiss Berl Akademie-Verlag, Berlin. **2**: 177-182.

Briggs, D.E. (1978). Barley Genetics. In: Barley (Ed. Briggs, D.E.), London: Chapman and Hall Ltd, pp 419-441.

Epstein, E., Norlyn, J.D., Rush, D.W., Kingsbury, R.W. and Kelley, D.B. (1980). Saline culture of crops: a genetic approach. Science **210**: 399-404.

Farquhar, G.D. and Richards, R.A. (1984). Isotopic composition of plant carbon correlates with water-use efficiency of wheat genotypes. Australian Journal of Plant Physiology **11**: 539-552.

Forster, B.P., Pakniyat. H., Macaulay, M., Matheson, W., Phillips, M.S., Thomas, W.T.B. and Powell, W. (1994). Variation in the leaf sodium content of *Hordeum vulgare* (barley) cultivar Maythorpe and its derived mutant cv. Golden Promise. Heredity **73**: 249-253.

Franckowiak, J.D. (1991). Allelism tests among selected semi-dwarf barleys. Barley Genetics Newsletter **21**: 17-23.

Gustafsson, A., Hagberg, A., Persson, G. and Wiklund, K. (1971). Induced mutations and barley improvement. Theoretical and Applied Genetics **41**: 239-248.

Haahr, V. and von Wettstein, D. (1976). Studies on an induced high yielding dwarf mutant of spring barley. In: Barley Genetics III, Proceedings of the Third International Barley Genetics Symposium, 1975 (Ed. Gaul, H.). Verlag Karl Thiemig, Munich, Germany, pp. 215-218.

Hanson, P.R., McVittie, J.A. and Smalley, J.L. (1980). Seedling response to exogenous gibberellic acid in spring barley. Zeitschrift für Planzenzüchtung **84**: 115-121.

Jacobsen, T. and Adams, R.M. (1968). Salt and silt in ancient Mesopotamian agriculture. Science **128**: 1251-1258.

Kapala, A. and Rybinski, W. (1995). Variability of seed protein composition of hulless mutants of spring barely (*Hordeum vulgare* L.). Journal of Applied Genetics **36**: 129-135.

Kucera, L., Lundqvist, U. and Gustafsson, A. (1975). Induction of breviaristatum mutants in barley. Hereditas **80**: 263-278.

Kumar, A. (1996). The adventures of the TY1-copia group of retrotransposons in plants. Trends in Genetics **12**: 41-43.

Laurie, D.A., Pratchett, N., Romero, C., Simpson, E. and Snape J.W. (1993). Assignment of the *denso* dwarfing gene to the long arm of chromosome 3 (3H) by use of RFLP markers. Plant Breeding **111**: 198-203.

Lundqvists, U. (1986). Barley mutants – diversity and genetics. In: Svalöf 1886-1986, Research and Results in Plant Breeding (Ed. Olsson, G.) Svalöf AB 268 00, Svalöv, Sweden, pp85-88.

Müller, H.J. (1927). Artificial transmutation of the gene. Science **66**: 84-87.

Pakniyat, H., Baird, E., Thomas, W.T.B., Caligari, P.D.S., Powell, W. and Forster, B.P. (1996). Effect of semi-dwarf mutants on salt tolerance in barley. In: Slinkard, A.E., Scoles, G.J. and Rossnagel, B.G (eds). Proceedings of Fifth International Oat Conference and Seventh International Barley Genetics Symposium, Saskatoon 1996. Pp. 660-661. University of Saskatchewan, Saskatoon.

Pakniyat, H.J. (1996). Genetic studies on salt tolerance in barley (*Hordeum vulgare* L.). PhD thesis. The University of Reading, Reading, UK.

Pakniyat, H, Handley, L.L., Thomas, W.T.B., Connolly, T., Macaulay, M., Caligari, P.D.S. and Forster, B.P. (1997a). Comparison of shoot dry weight, Na^+ content and $\delta^{13}C$ values of *ari-e* and other semi-dwarf barley mutants under salt stress. Euphytica **94**: 7-14.

Pakniyat, H., Thomas, W.T.B., M.,Caligari, P.D.S. and Forster, B.P. (1997b). Comparison of salt tolerance of *GPert* and non-*GPert* barleys. Plant Breeding **116**: 189-191.

Pearce, F. (1987). Banishing the salt of the earth. New Scientist **11**: 53-56.

Persson, G. (1969). An attempt to find suitable genetic markers for dense ear loci in barley. Hereditas **62**: 25-96.

Russell, J., Fuller, J., Young, G., Thomas, W.T.B., Taramino, G., Macaulay, M., Waugh, R. and Powell, W. (1997). Discriminating between barley genotypes using microsatellite markers. Genome **40**: 442-450.

SCOPE (1978). Scientific Committee on Problems of Environment. In: Land irrigation in developing countries (ed. Worthington, E.B.), Pergamon Press, Oxford, pp3-8.

Sigburbjorsson, B. and Micke, A. (1969). Progress in mutation breeding. In: Induced Mutation in Plants. Proceedings of an International Symposium on: Nature, Induction and Utilization of Mutation in Plants. International Atomic Energy Agency, Vienna, 1969, 673-697.

Stadler, L.J. (1930). Some genetic effects of X-rays in plants. Journal of Heredity **21**: 3-19.

Stoner, R. (1988). Engineering a solution to the problem of salt laden soils. New Scientist **3**: 44-46.

Thomas, W.T.B., Asher, M.J.C., Swanston, J.S. and Thomas, C.E. (1988). The transfer of resistance to powdery mildew from *Hordeum spontaneum* to the spring barley cv. Golden Promise. In: Jorna, M.L. and Slootmaker, L.A. (eds), Cereal Breeding Related to Integrated Cereal Production, Proc. EUCARPIA Conf., pp. 200-204. Pudoc, Wageningen.

Thomas, W.T.B., Powell, W. and Wood, W. (1984). The chromosomal location of the dwarfing gene present in the spring barley variety Golden Promise. Heredity **53**: 177-183.

Vos, P., Hogers, R., Bleeker, M., Reijans, M., Lee, T.V.D., Hornes, M., Frijters, A., Pot, J., Peleman, J., Kuiper, M. and Zabeau, M. (1995). AFLP: a new technique for DNA fingerprinting. Nucleic Acids Research **23**: 4407-4414.

Wan, Y. and Lemaux, P.G. (1994). Generation of large numbers of independently transformed fertile barley plants. Plant Physiology **104**: 37-48.

Zabeau, M. and Vos, P. (1993). Selective restriction fragment amplification: a general method for DNA fingerprinting. European Patent Application No: 92402629.7, Publication No: 0534858.

Zwar, J.A. and Chandler, P.M. (1995). α-Amylase production and leaf protein synthesis in a gibberellin-responsive dwarf mutant of 'Himalay' barley (*Hordeum vulgare* L.). Planta **197**: 39-48.

Mapping genes for flowering time and frost tolerance in cereals using precise genetic stocks

J.W. Snape[1], R. Sarma[1], S.A. Quarrie[1], G. Galiba[2] and J. Sutka[2]

[1] *John Innes Centre, Norwich Research Park, Colney, Norwich, NR4 7UJ, UK*

[2] *Agricultural Research Institute of the Hungarian Academy of Sciences, 2462 Martonvasar, Hungary*

Key words: comparative mapping, frost tolerance, *Triticum aestivum*, vernalization

Summary

The development of comprehensive genetic maps based on molecular markers has increased the power of genetical analysis immensely in the last few years. Characters previously recalcitrant to analysis, such as abiotic stress responses, are now amenable, and individual major genes and QTL mediating the variation can be identified. This has allowed the development of strategies for stress amelioration by adjusting the timing of the life cycle and introducing genes which enable the plant to tolerate stress. These strategies are illustrated here by showing how genes for vernalization response and cold tolerance on chromosomes 5A and 5D of wheat have been identified and located. Additionally, their relationships to genes in other species, such as barley and rice can be characterised through comparative mapping approaches, leading to strategies for their isolation using rice genomic tools.

M. Maluszynski and K.J. Kasha (eds.),
Mutations, In Vitro and Molecular Techniques for Environmentally Sustainable Crop Improvement, 17–25.

Introduction

Abiotic stresses are a major barrier to sustainable crop production in temperate and tropical environments, and the yield potential of any particular cereal variety is rarely maximised in any one season or area, due to the limitations imposed by abiotic stresses. World-wide, there is the potential to encounter a whole range of stresses such as extremes of temperature, drought, nutrient deficiencies, soil contamination and pollution of soil or atmosphere by organic chemicals. To date, little is known about the genetics of cereal responses to abiotic stress. It is recognised, however, that varieties do differ in their responses to different environmental conditions and, thus, that differences are genetically based. The challenge is to identify sources of genetic variation for tolerance to important stresses, and then to develop strategies for their deployment so that plant breeders can develop new cultivars with greater yield potential and yield stability over seasons, and eco-geographic locations. There are two ways of achieving this 1) by understanding the genetics of adaptation one can adjust the timing of the life cycle, or 2) by identifying and introducing genes which allow the plant to tolerate the stress environment.

Clearly, the timing of the components of the life cycle determines how and when the plant develops through the juvenile stages, when flowering is initiated, when flowering takes place, and when grain filling and ripening occur. Appropriate adjustments to these phases can avoid winter damage from cold temperatures, avoid cool temperatures at flowering that result in male sterility, and escape drought and high temperatures in the warmer summer months. By understanding the genetic control of the duration of these phases, new varieties can be designed and bred to avoid or minimise the stress risk. The genetic control of flowering time in cereals is complex, and chromosome substitution line analysis in wheat has shown that chromosomes of nearly all homoeologous groups are involved (Law *et al.*, 1991). By homology, this would also be expected to be the case in other temperate cereals such as barley, oats and rye. The existence of extensive allelic variation for these genes is probably the basis of the wide adaptation of cereals, world-wide. Notwithstanding, single major genes can be identified, and the variation identified can be classified into the effects of three separate sets of genes, those controlling vernalization, those controlling photoperiod response and, those controlling developmental rate (independent of vernalization and photoperiod), so called 'earliness *per se*' genes. With respect to yield potential, studies also show that genes of all three systems have pleiotropic effects on other aspects of plant growth and development which have important consequences for cereal breeding for specific adaptation. Comparative genetic studies also show that these genes have been conserved through evolution, so that studies in one species has relevance to studies of the other cereals (Moore *et al.*, 1995). This is particularly the case in the Triticeae, where genes for all three flowering responses can be recognised as being syntenic. Our current knowledge of the location of flowering time genes in wheat and barley, for example, is shown in Table 1.

The genetic control of stress responses is equally complex, and most variation seems to be mediated by QTL. Nevertheless, some genes of major effect have been identified, such as genes controlling responses to drought (Quarrie *et al.*, 1995). However, because of the complexities of measuring stress responses at the phenotypic, physiological and biochemical levels much less is presently known about the genetical control of stress responses than flowering time and other more highly heritable traits.

The location of genes controlling these complex traits is now possible by the application of marker-mediated techniques. Such modern methods of genetical analysis can now lead to the location of major genes and QTL mediating flowering time and tolerance of abiotic stresses in cereals. This is achieved by exploiting precise genetic stocks, such as doubled haploid and recombinant substitution lines, along with the comprehensive genetic maps now available through the application of molecular marker techniques. Cross-hybridizing RFLP probes allow a comparative genetics approach, so that information obtained in one species can be translated across species to aid the detection, identification, tagging and, ultimately, the cloning of flowering time and stress resistance genes. Additionally, the co-location of genes will allow the evaluation of whether different alleles for flowering time and stress responses are pleiotropic effects of the same genes, or form adapted gene complexes of genes co-selected by natural selection for adaptation. In our studies major efforts have been made to map genes controlling flowering time so that stress-avoidance strategies can be developed, and also mapping genes, which directly contribute to stress tolerance through mechanisms which protect the plant at the cellular level.

Table 1. The chromosomal locations of genes controlling flowering time that have been intra-chromosomally mapped in wheat and barley

	A Genome	B Genome	D Genome	H Genome*
Group 1				*Ppd-H2* *Vrn-H3 (Sh3)*
Group 2	*Ppd-A1*	*Ppd-B1* *Eps-2BS*	*Ppd-D1* *Eps-2DS*	*Ppd-H1* *Eps-2HS*
Group 3				*Eps-3HL*
Group 4				*Vrn-H2 (Sh)* *Eps-4HL*
Group 5	*Vrn-A1* *Vrn-A2 (4A)*	*Vrn-B1*	*Vrn-D1*	*Vrn-H1 (Sh2)* *Eps-5HL*
Group 6				*Eps-6HL.1* *Eps-6HL.2*
Group 7		*Vrn-B4*		*Eps-7HS* *Eps-7HL*

**From Laurie et al., 1995 and Stracke and Borner, 1998*

The genetical approach to mapping genes for flowering time and stress responses

Marker-mediated genetical analysis relies on the establishment of an association between the segregation of known marker alleles with differences in phenotypic expression of the trait. This has four steps.

1) It requires the development of an appropriate recombinant population between parents differing in phenotype for the character(s) of interest. This can either be a F_2 or F_3, or preferably, recombinant inbred lines derived from single seed descent or doubled haploid procedures. In certain species, such as wheat, specialist lines can be developed by chromosome manipulation, such as recombinant substitution lines.
2) A genetic map of the recombinant population is developed by characterizing all individuals or lines for their allelic constitution at molecular marker loci dispersed

throughout the genome and identified as being polymorphic in the cross from examination of the parental lines. These markers can be either RFLP or PCR based, although increasingly the approach has been to use widely spaced RFLP markers to 'anchor' the map and then to use more efficient PCR methods such as Amplified Fragment Length Polymorphisms (AFLP) or Simple Sequence Repeats (SSR - also known as microsatellites) to fill the gaps between them. Probes or primers are chosen through reference to published maps and the availability of particular probes or sequences. From the segregation data the genetic map can be developed using software such as MapMaker (Lander, 1993) or JoinMap (Stam, 1993).

3) Concurrent with map development will be the evaluation of the recombinant lines in appropriate replicated and randomised experiments for the traits of interest.
4) The final step is to use statistical procedures to combine the map and trait data so as to partition the genotypic variation into effects of individual QTL and to locate these relative to the marker loci in defined chromosome segments. A number of statistical methods are now available to enable this to be done from simple ANOVA and regression techniques through to more sophisticated maximum likelihood and least squares approaches. Computer software is available to implement these procedures such as QTL Cartographer (Zeng, 1994), MapMaker/QTL (Lander and Botstein, 1989) and JaqcMap (http://www.sun1.bham.ac.uk/g.g.seaton).

Different cereals species have different advantages for genetical analysis. For example, barley is a diploid which has facile doubled haploid techniques, high levels of molecular marker polymorphisms, a large range of RFLP probes, and recently, several hundred SSR primer sets. Hence, it is now fairly routine to develop high resolution maps in populations of interest in barley for QTL mapping. We have used such an approach to characterize the complete genetical control of flowering time differences between European winter and spring barleys (Laurie *et al.*, 1995), as shown in Table 1. In this analysis it was possible to show that variation for alleles at twelve loci is, probably, responsible for most of the wide variation in flowering time in barley. In wheat, the most efficient method of genetical analysis for QTL is to exploit the large range of aneuploid derived stocks available, particularly single chromosome substitution lines. By cytogenetical based chromosome manipulation techniques, single chromosomes can be transferred systematically from one variety to another, and hence genetical differences, mediated by single chromosomes, can be examined on a homogenous genetic background (Law *et al.*, 1987). From backcrossing a substitution line with its recipient variety, recombinant lines for the single chromosome (recombinant substitution lines) can be developed for genetic analysis using the technique described by Law (1966). By the use of these lines the precise location of QTL can be obtained because the background genetic variation is effectively eliminated and the trait variation is only influenced by alleles segregating between the single pair of parental chromosomes and associated environmental effects. We have used these lines as the basis for the location of genes affecting flowering time and stress responses in wheat, particularly concentrating on chromosomes of homoeologous Group 5 (5A and 5D). We have concentrated on the location of genes on these chromosomes since our previous studies have shown that these carry major determinants of flowering time, particularly vernalization response and genes controlling a range of stress responses such as tolerance of freezing temperatures, drought, osmotic stress and high temperatures.

Mapping genes for vernalization requirement and stress responses to 5A and 5D of wheat

Major genes controlling sensitivity to vernalization, the *Vrn* genes, determine the control of the spring wheat/winter wheat growth habit differences. Winter types require vernalization to initiate flowering and spring types do not. The lack of vernalization requirement is generally dominant. Five loci are known to control spring/winter differences and the chromosomal locations of four of them have been established using chromosome substitution line analysis, namely *Vrn-A1* (on chromosome 5A), *Vrn-D1* (on chromosome 5D), *Vrn- B1* (on chromosome 5B) and *Vrn-B4* (on chromosome 7B) (Law *et al.*, 1976; Law *et al.*, 1991). In the major winter wheat growing areas of the world, studies have shown that alleles at the *Vrn-A1* locus appear to be predominant in reducing the vernalization requirement (Pugsley, 1971; Snape *et al.*, 1985). Thus by adjusting the allele at *Vrn-A1*, it is possible to design wheats which are winter, spring or facultative, (which can be winter or spring sown). For example, Snape *et al.* (1985) have shown that it is possible to convert a UK winter wheat into a spring wheat by transferring a vernalization insensitive allele at this locus, with no disadvantageous effects on other traits from late autumn or spring sowing times. However, with the trend towards earlier sowing in many areas of the world, it is necessary to have a vernalization requirement if an early sowing is desired, as floral initiation will otherwise occur before winter, with concomitant risks of frost damage. Adjustment of vernalization requirement can, therefore, tailor varieties to different times of sowing and thus maximise their yield potential.

In addition, although studies have shown that the genetic control of frost resistance is complex and can be regarded as a polygenic trait (Sutka, 1981), major genes have been mapped on chromosomes 5A and 5D using molecular markers (Galiba *et al.*, 1995; Snape *et al.*, 1998). The major question is, therefore, whether vernalization response and cold tolerance are pleiotropic effects of the *Vrn* genes or determined by separate loci. To answer this question requires precise mapping of the loci involved in both traits, using a marker-mediated approach.

To map genes affecting flowering time and cold tolerance on chromosome 5A, a population of single chromosome recombinant lines, derived from the cross between the single chromosome substitution lines Chinese Spring (*Triticum spelta* 5A), which is vernalization insensitive, and Chinese Spring (Cheyenne 5A) which is vernalization sensitive, was developed. The flowering time of each line was characterised in a growth chamber under a regime of 16 h day - 8 h night at temperatures of 22 and 15°C, respectively, without prior vernalization. Frost tolerance was assessed using low temperature treatments varying from -10 to -15° C, following hardening. Discontinuous variation for flowering time and frost responses was observed when the phenotypes were assessed under these appropriate conditions. RFLP mapping techniques were applied to the lines to develop and anchor the genetic map, and this showed that *Vrn-A1* and *Fr1*, the locus for cold tolerance, were separate, but closely linked loci, on the long arm of chromosome 5A. Both loci were closely linked to the RFLP markers, *Xpsr2021(ABA2)* and *Xpsr426* distal on the long arm, but proximal to the 5A/4A translocation, Figure 1.

Recently, we have shown that the presence of Cheyenne alleles for vernalization requirement and frost tolerance on chromosome 5A, significantly increased sugar concentrations, particularly sucrose and fructan, and hence improve osmotic potential and frost tolerance (Galiba *et al.*, 1997). This confirms that these genes may contribute to general stress tolerance and are an addition to or pleiotropic effects of, genes responsible for frost tolerance, drought tolerance and osmotic stress previously detected in this region through mechanisms which act at the cellular level.

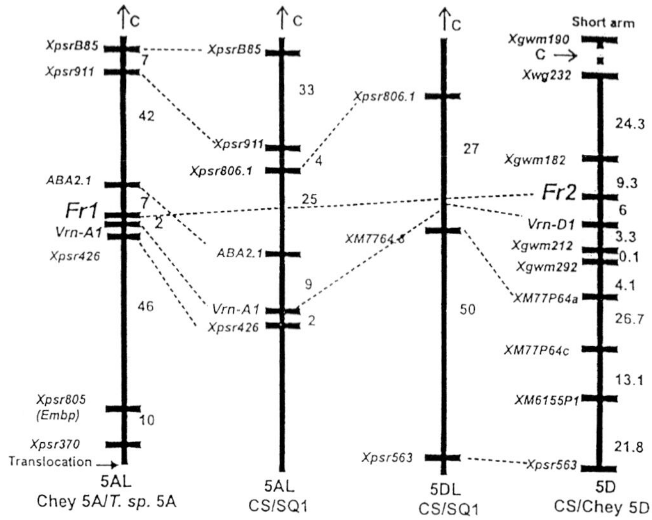

Figure 1. Genetic map of chromosome 5D from the CS x CS (Cheyenne 5D) cross showing the location of *Vrn-D1* and *Fr2*, and comparative map with chromosomes 5A and 5D of the CS x SQ1 cross, and 5A of the CS (*T.spelta* 5A) x CS (Cheyenne 5A) cross, showing positions and homoeologies of *Vrn-D1* with *Vrn-A1* and *Fr2* with *Fr1*

The analysis of chromosome 5A has been extended to examine if there are homoeologous loci for vernalization response and stress responses, and particularly frost tolerance, on chromosome 5D which previous evidence had shown to contain the vernalization gene *Vrn-D1* (formerly *Vrn3*). A recombinant substitution line population was developed from the cross between Chinese Spring, which is vernalization insensitive, and the single chromosome substitution line Chinese Spring (Cheyenne 5D), which is vernalization sensitive. Since previous work has shown that chromosome 5D has few RFLP, the strategy used for mapping was to use the RFLP probes previously located on the homoeologous group 5 chromosomes to search for polymorphisms to provide 'anchor markers' for the map and then to use Amplified Fragment Length Polymorphism (AFLP) and microsatellite markers to provide greater resolution. Each line was characterised phenotypically for flowering time and frost tolerance as described above. The data for flowering time revealed a bimodal segregation based around the parental scores, indicating the segregation of a single gene, presumed to be *Vrn-D1*. Although there was also a suggestion of bimodality for frost responses, it was not possible to unambiguously characterize the genotype of each individual line into susceptible and tolerant classes, and hence QTL analysis had to be used to locate the gene(s) for frost tolerance.

Vrn-D1 was located, as expected, distally on the long arm of 5D, closely linked to the microsatellite markers Xgwm212 and Xgwm292. QTL analysis for cold response revealed that *Vrn-D1* and markers closely linked showed a strong association with the QTL for frost tolerance. This indicated the presence of the predicted locus, *QFr.jic-5D*. However the map position suggests that *QFr.jic-5D* is a linked locus and not a pleiotropic effect of *Vrn-D1*, since the maximum likelihood position was about 6cM proximal to *Vrn-D1*.

The individual mapping data for the 5A and 5D chromosomes shows that the *Vrn-A1* and *Vrn-D1* loci both map at distal locations on homoeologous long arms, and are thus, probably homoeoallelic. To prove this further, the maps were aligned using cross hybridizing probes and 'bridging' maps of 5A and 5D developed on a doubled haploid population from the cross of Chinese Spring x SQ1 (Quarrie *et al.*, 1995). This revealed that *Vrn-A1* and *Vrn-D1* are, in all probability, homoeoallelic, as are *Fr1* and Q*Fr.jic-5D*. These loci are thus probably derived from the common ancestor of the A and D genomes and predate the speciation of bread wheat. From other studies, the *Vrn* and *Fr* genes on the homoeologous group 5 chromosomes show conserved synteny with barley (chromosome 7, 5H) (Laurie *et al.*, 1995), rye (5R) (Plaschke *et al.*, 1993), and probably oats linkage group 24 (Sarma *et al.*, 1998). Thus, there are likely to be genes mediating stress tolerance on these chromosomes as well. A suggestion that this is the case comes from the studies of Pan *et al.* (1994) who located a QTL for frost tolerance on barley chromosome 7 (5H), in a homoeologous location.

Physical mapping of stress tolerance genes and comparative genetical analysis with rice

Both wheat and barley precise genetic stocks are excellent tools for the identification and mapping of major genes and QTL. However, both of these species lack the genomic tools, such as large insert libraries, to enable genes for flowering time and stress responses to be cloned and, hence, studied at the molecular level. This can, however, be achieved by using comparative mapping and rice genomic tools, where comparative mapping with rice using cross-hybridizing RFLP probes is highly developed for the grasses since Moore *et al.* (1995) have shown how all the cereal genomes can be aligned in terms of rice linkage blocks. If this is to be carried out for *Vrn, Fr* and other of the linked stress response genes, it is necessary to first determine the homoeology of the *Vrn/Fr* region of the Triticeae with rice. By using rice probes supplied by the Japanese Rice Genome Program and using a mapping population developed at the John Innes Centre between the *indica* variety IR20 and the *javanica* variety 63-83 it was shown that a region homoeologous to the Triticeae *Vrn* region exists on rice chromosome 3. This finding was confirmed using deletion lines for chromosome 5A, developed in the variety Chinese Spring (Gill *et al.*, 1996), where probes from rice chromosome 3 and probes co-segregating with *Vrn-A1* all mapped between deletion breakpoints 0.68 and 0.78. Intriguingly this region in rice contains a flowering time QTL, FLTQ1, mapped by Sarma *et al.* (1998) in the marker intervals *psr145* to *R3226*. Moreover, other markers (*R2404*, *R2311*) associated with *Hd-6,* another flowering time QTL mapped onto the long arm of rice chromosome 3 by Yamamoto *et al.* (1996) have been mapped onto the same chromosomal region of the deletion lines flanked by breakpoints 0.68 and 0.78. However, it cannot be determined if these QTL, FTLQ1 and Hd-6, are the same or different loci with present data. Nevertheless, these data suggest that rice may have active *Vrn* homologues.

Conclusions

The types of genetical analyses described above are at last revealing detailed information about the genetical control of adaptation and stress responses in wheat and the influence of specific genes on yield performance in different environments. The challenge now is to translate this genetical information into selection tools that can be used by the wheat breeders on an efficient and low cost basis. This will require the development of PCR-based gene tags closely linked to the genes of interest so that a high through-put of material in segregating generations of breeders nurseries is possible. Also it will enable plant breeders to pyramid desirable genes from different sources, and to be able to do this at juvenile stages of growth by extracting DNA from young plants, and without the necessity for creating artificial environments to mimic the stress environment.

The studies have revealed that there is a region of chromosomes of homoeologous group 5 of the *Triticeae* and chromosome 3 of rice, and probably, by homology, in all grasses, that is very important for eco-geographical adaptation. These regions carry genes affecting the timing of the life cycle and responses to environmental stress. The tagging of this region using molecular markers will enable plant breeders to carry out marker-assisted selection for alleles giving improved stress resistance. Additionally, developments in functional genomics may allow the gene contents of such regions to be elucidated in detail. Currently, efforts are beginning to sequence the rice genome through an international collaboration. Eventually, when the region of current interest is sequenced it will be possible to reveal how many genes are involved in these responses. The challenge will then be to translate the sequence information into the mechanisms of action to enable further genetic modification of the genes and their re-insertion into cereals in a directed manner.

References

Galiba, G., I. Kerepsi, J.W. Snape and J. Sutka, 1997. Location of a gene regulating cold induced carbohydrate production on chromosome 5A of wheat. Theor.Appl.Genet. **95**: 265-270

Galiba, G., S.A. Quarrie, J. Sutka, A. Morgounov and J.W. Snape, 1995. RFLP mapping of the vernalization and frost tolerance genes on chromosome 5A of wheat. Theor.Appl.Genet. **90** : 1174-1179

Gill, K.S., B.S. Gill, T.R. Endo and E. Boyko, 1996. Identification and high-density mapping of gene-rich regions in chromosome group 5 in wheat. Genetics. **143**: 1001-1012

Lander, E.S. (1993) MapMaker/Exp 3.0 and MapMaker/QTL 1.1. Whitehead Institute, Cambridge MA

Lander, E.S. and D. Botstein, 1989. Mapping Medelian factors underlying quantitative traits using RFLP linkange maps. Genetics. **121**: 185-199

Laurie, D.A., N. Pratchett, J. Bezant and J.W. Snape, 1995. RFLP mapping of five major genes and eight QTLs controlling flowering time in a winter x spring barley cross. Genome. **38** : 575-585

Law, C.N., 1966. The locations of genetic factors affecting a quantitative character in wheat. Genetics. **53**: 487-498

Law, C.N., C. Dean and G. Coupland, 1991. Genes controlling flowering and strategies for their isolation and characterisation. In: The Molecular Biology of Flowering. Jordan, B. (Ed.) CAB International, Oxford, pp.47-68

Law, C.N., J.W. Snape and A.J. Worland, 1987. Aneuploidy in wheat and its use in genetic analysis. In: Wheat Breeding: Its Scientific Basis. Lupton, F.G.H. (Ed.) Chapman and Hall, London. pp.71-108

Law, C.N., A.J. Worland and B. Giorgi, 1976. The genetic control of ear-emergence time by chromosomes 5A and 5D of wheat. Heredity. **36**: 49-58

Moore, G., K.M. Devos, Z. Wang and M.D. Gale, 1995. Grasses, line up and form a circle. Current Biology. **5**(7): 737-739

Pan, A., P.M. Hayes, F. Chen, T.H.H. Chen, T. Blake, S. Wright, I. Karsai and Z. Bedo, 1994. Genetic analysis of the components of winter hardiness in barley (*Hordeum vulgare* L.). Theor.Appl.Genet. **89**: 900-910

Plaschke, J., A. Börner, D.X. Xie, R.M.D. Koebner, R. Schlegel and M.D. Gale, 1993. RFLP mapping of genes affecting plant height and growth habit in rye. Theor.Appl.Genet. **85**: 1049-1054

Pugsley, A.T., 1971. A genetic analysis of the spring-winter habit in wheat. Aust.J.Agric.Res. **22**: 21-31

Quarrie, S.A., G. Galiba, J. Sutka, J.W. Snape, A. Semikhodskii, A. Steed, M. Gulli and D.C. Calestani, 1995. Association of a major vernalization gene of wheat with stress-induced abscisic acid production. In: Crop Adaptation to Cool Climates. Proc. COST 814 Workshop, Hamburg, Oct.1994. Dorffling, K., B. Brettschneider, H. Tatau and K. Pitham (Eds.) European Commission, Brussels. pp.403-414

Sarma, R.N., B.S. Gill, T. Sasaki, G. Galiba, J. Sutka, D.A. Laurie and J.W. Snape, 1998. Comparative mapping of the wheat chromosome 5A *Vern-A1* region with rice and its relationship to QTL for flowering time. Theor.Appl.Genet. **97**: 103-109

Snape, J.W., C.N. Law, B.B. Parker and A.J. Worland, 1985. Genetical analysis of chromosome 5A of wheat and its influence on important agronomic characters. Theor.Appl.Genet. **71**: 518-526

Snape, J.W., A. Semikhodskii, R.N. Sarma, V. Korzun, L. Fish, S.A. Quarrie, B.S. Gill, T. Sasaki, G. Galiba and J. Sutka, 1998. Mapping vernalization loci in wheat and comparative mapping with other cereals. In: Proc. 9th International Wheat Genetics Symposium Vol.3. Sliknard, A.E. (Ed), University Extension Press, University of Saskatchewan. pp. 156-158

Stam, P., 1993. Construction of integrated genetic linkage maps by means of a new computer package: JoinMap. Plant J. **5**: 739-744

Stracke, S. and A. Borner, 1998. Molecular mapping of the photoperiod response gene ea_7 in barley. Theor. Appl. Genet. **97**: 797-800

Sutka, J., 1981. Genetic studies of frost resistance in wheat. Theor.Appl.Genet. **59**: 145-152

Yamamoto, T., Y. Kuboka, S.Y. Lin, T. Shimano, A. Shomura, M. Yano and T. Sasaki, 1996. Fine mapping and characterization of quantitative trait loci for heading time in rice. In: Plant Genome IV. Scherago International, New York. pp.57 (Abstract)

Zeng, Z.B., 1994. Precision mapping of quantitative trait loci. Genetics. **136**: 1457-1468.

Characterization of wheat-*Thinopyrum* partial amphiploids for resistance to barley yellow dwarf virus

G. Fedak[1], Q. Chen[2], A. Comeau[3] and C.A. St.-Pierre[4]
[1] *Eastern Cereal and Oilseed Research Centre, Agriculture and Agri-Food Canada, Ottawa, Ontario, K1A 0C6 (on staff at Cereal Research Centre, Winnipeg)*
[2] *Research Centre, Agriculture and Agri-Food Canada, P.O. Box 3000, Lethbridge, AB, T1J 4B1*
[3] *Centre de Recherche, Agriculture and Agri-Food Canada, St.-Foy, QC, G1V 2J3*
[4] *Université Laval, Dept. De Phytologie, Quebec, QC, G1K 7P4*

Key words: amphiploids, GISH, *Thinopyrum*, virus resistance, wheat

Summary

Resistance to viruses such as wheat streak mosaic virus (WSMV) and barley yellow dwarf virus (BYDV) is lacking in the primary gene pool of wheat and therefore is being introgressed from wild relatives such as *Thinopyrum* species. Resistance to BYDV was found in partial amphiploids ($2n = 8x = 56$, consisting of 42 wheat and 14 alien chromosomes) obtained in hybrids between wheat and both *Th. intermedium* and *Th. ponticum*. GISH analysis revealed that the alien genomes of all but one resistant partial amphiploid were heterogeneous consisting of different ratios of St, J^s and J genome chromosomes obtained from the *Thinopyrum* parent. Translocated chromosomes consisting of Robertsonian, interstitial and terminal translocations between the different genomes were also detected. The tissue blot immunoassay showed that partial amphiploids having resistance could be inoculated with the virus but both virus multiplication and spread were completely blocked.

M. Maluszynski and K.J. Kasha (eds.),
Mutations, In Vitro and Molecular Techniques for Environmentally Sustainable Crop Improvement, 27–34.

Introduction

The wild wheatgrasses *Thinopyrum intermedium* (Host) Barkworth and Dewey ($2n = 6x = 42$) and *Th. ponticum* (Podp.) Barkworth and Dewey ($2n = 10x = 70$) have made major contributions to increasing the frequency of disease resistant alleles in the gene pool of wheat. It was shown that approximately 30 alleles in total have been transferred into wheat from these two species (Fedak, in preparation). This total includes several alleles for resistance to each of leaf rust, stem rust, BYDV, yellow rust and WSMV. The two species not only carry immunity to many pathogens, but are readily hybridized with wheat and backcross progenies are easily obtained.

There are several designations for the genomes of *Thinopyrum intermedium* (6*x*) but the most recent, based on fluorescent *in situ* hybridization (FISH) analysis designates the three genomes as JJJ^sJ^s St St and St (Chen *et al.*, 1998c) and those for *Thinopyrum ponticum* as $JJJJJJ^sJ^sJ^sJ^s$.

Barley yellow dwarf virus is one of the most important virus diseases in the world (Comeau and Dubuc, 1977; Gill, 1980; Cisar *et al.*, 1982). There is no resistance to this virus in the primary or secondary gene pools of wheat, (although the Brazilian wheat cultivar Maringa contains a useable form of resistance), therefore resistance must be obtained from the tertiary gene pool. Certain accessions of both the above species are immune to the virus as are their hybrids with wheat, amphiploids, partial amphiploids and certain critical wheat-*Thinopyrum* chromosome addition lines. This report will describe the characterization of the virus resistance in wheat-*Thinopyrum* partial amphiploids.

Materials and Methods

The various groups of cytogenetic stocks were analyzed by one of three methods: two for virus assay and one for genome constitution:

a. slot blot hybridization using a partial DNA clone of the coat protein gene of the PAV strain of the virus
b. tissue blot immunoassay
c. genomic *in situ* hybridization (GISH).

The partial amphiploids reported herein and their origins are detailed in Table 1. Partial amphiploids arise as backcross 1 progenies following the backcrossing of wheat onto hybrids between wheat and polyploid *Thinopyrum* species. The genomes in such aneuploids consist of 42 wheat and 14 *Thinopyrum* chromosomes.

The partial amphiploids analyzed by slot blot hybridization are listed in Table 1. The method briefly involves the inoculation of seedlings at the 3-4 leaf stage by viruliferous aphids, extracting the viral RNA and blotting onto a positively charged nylon membrane and probing with a radioactively labeled coat protein gene as described by Zhang *et al.* (1996).

The plant materials analyzed by the tissue blot immunoassay (TBIA) are listed in Table 2. Briefly, the method involves inoculation of seedlings at the 3-4 leaf stage with viraliferous aphids, killing off the aphids, cutting the test leaves and stems with a razor blade and blotting the cut ends onto a nitrocellulose membrane. This membrane is then probed with a BYDV polyclonal antiserum and stained with nitro blue tetrazolium. The method is described in detail by Chen *et al.* (1997). The methods for GISH analysis were followed as described by Zhang *et al.* (1996). The genome symbols as proposed by Wang *et al.*(1994) have been employed in this manuscript.

Results and discussion

The reaction of 17 partial amphiploids to BYDV serotype PAV is shown in Table 1. Partial amphiploids 693, 7631, 7430, 68 and 784 were produced from wheat x *Thinopyrum ponticum* (Li *et al.*, 1985). They were derived from different wheat parents crossed with the same accession of *Thinopyrum ponticum*. Of the five partial amphiploids, only line 784 was immune while the other four lines were sensitive to the PAV strain of BYDV. The partial amphiploids 40767-1 and 40767-2 were obtained from another common BYDV resistant source of *Thinopyrum ponticum*. The former line was uniformly susceptible while the latter contained both immune and highly resistant individuals in the same population. OK7211542, developed from a hybrid between wheat and *Thinopyrum ponticum* proved to be one of the most resistant partial amphiploids derived from that source (Chen *et al.*, 1997; Chen *et al.*, 1998b).

The Zhong series of partial amphiploids listed in Table 1 plus strain 78829 were all derived from the same accession of *Thinopyrum intermedium*. Partial amphiploids Zhong 3, 4, 5, 6 & 7 and strain 78829 all proved to be highly resistant to BYDV by the slot blot test. TAFA6, isolated by Cauderon *et al.* (1973) from backcrossing a wheat x *Thinopyrum intermedium* hybrid with wheat and shown to have resistance to rust fungal pathogens was also highly resistant to BYDV. In all cases of resistance, the plants could be inoculated with the virus but viral spread was inhibited.

Though not shown in Table 1, another 9 partial amphiploids derived from wheat x *Thinopyrum junceum* and 13 disomic addition lines derived from these sources were tested for BYDV resistance by the same means. None showed any level of resistance indicating that the *Thinopyrum junceum* accession used as a parent probably was not resistant to BYDV.

The BYDV resistant partial amphiploids shown in Table 1 were analyzed by genomic *in situ* hybridization (GISH) to determine their genomic constitution. Of the eleven partial amphiploids analyzed, only accession number 78829 had a complete alien genome; in this case the St genome derived from *Thinopyrum intermedium*. The alien genomes in all other partial amphiploids were heterogeneous i.e. composed of chromosomes from more than one parental genome. Derivatives of *Thinopyrum ponticum* had combinations of J^s and J genomes while *Thinopyrum intermedium* derivatives mostly had combinations of St and J^s genomes (Fig. 1). In addition, most alien genomes carried translocated chromosomes. The translocations usually involved St and J^s genome chromosomes and were either Robertsonian, intercalary or terminal interchanges (Fig. 1).

The partial hybrid alien genomes, although composed of two different constituent genomes likely contained chromosomes of all homoeologous groups. Forster *et al.* (1997) showed that six of seven addition lines produced from TAF46 all involved different homoeologus groups. The partial amphiploids that we studied were all perfectly fertile and had regular meiosis indicative of a balance of homoeologus chromosomes. However, we have shown that F_1 hybrids between different partial amphiploids had varying degrees of sterility caused by varying degrees of meiotic imbalance (Fedak *et al.*, 1999). This indicates that although alien genomes within any one accession are balanced for homoeologus chromosomes and translocations, they may differ from other partial amphiploids.

The genotypes used in the TBIA are shown in Table 2. Although the three partial amphiploids gave a resistant reaction in the slot blot test from field evaluation, it was suspected that firstly their levels of resistance were not all equal and secondly the mechanism of the resistance reaction was not known. The two wheat cultivars were tolerant to the virus i.e. susceptible to infection but suffered little yield loss and served as susceptible checks. The two barley strains both carried the yd_2 gene for BYDV resistance and served as resistant checks. The three categories of plant reaction were immune (0-1 phloem bundles stained) resistant (2-10 bundles stained) and susceptible (11-30 bundles stained).

Table 1. Reaction to BYDV serotype PAV of wheat-*Thinopyrum* partial amphiploids and karyotypes of resistant lines

Accession	Source	Reaction to BYDV-PAV	GISH karyotype of alien genome
a. Wheat x *T. ponticum* derivatives			
693	Li, *et al.* (1985)	S	-
7631	"	S	-
7430	"	S	-
68	"	S	-
784	"	R	6"J^s+1"J"; Zhang, *et al.* (1966)
40767-1	Zhang *et al.* (1996)	S	-
40767-2	Zhang *et al.* (1996)	R	4"J^s+1"J+1"J^s/1B+1J^s/E; Zhang, *et al.* (1996)
OK7211542	Sando 1930s (cited after Comeau *et al.*, 1994)	R	4"J^s+4"J; Chen, *et al.* (1998a)
b. Wheat x *T. intermedium* derivaives			
Zhong-1	Sun, 1981	S	-
Zhong-2	"	S	-
Zhong-3	"	R	5"St+1"St/J+1"St/J Zhang, *et al.* (1996)
Zhong-4	"	R	5"St+1"St/J+1"St/J
Zhong-5	"	R	2"St+2"J^s+3"St/J^s Chen, *et al.* (1998b)
Zhong-6	"	R	6"St+1"St/J
Zhong-7	"	R	5"St+1"St/J+1'St/J
78829	"	R	7"St (Zhang *et al.* 1996)
TAF46	Cauderon, *et al.* (1973)	R	3"St+4"J^s Chen *et al.* (1999)
OT Otrastayuskaya	V.I.R. St. Petersberg	S	4"St+5"S^s+1"W-J^s+1"W-St

* represents an St/J Roberstonian translocation; ** represents an St/J intercalary translocation

As shown in Table 2, all barley plants gave an immune reaction after viral infection while all wheat plants gave a susceptible reaction. All plants of partial amphiploids Zhong 4 and TAF46 gave a resistant reaction while 2 plants of OK7211542 gave a resistant reaction and 18 gave an immune reaction.

In order to determine whether the resistance of the OK line was an inhibition of entry of the virus into plant tissue or a restriction of systemic viral movement in infected tissues, a further test was carried out. A minicage containing viraliferous aphids was applied to 2 cm sections of the OK line, one barley line and wheat cultivar Pollet. At 6, 12, 18 and 24 days after inoculation, leaf sections and stem sections below the leaf were excised for TBIA testing. At 6 days after inoculation, virus particles were detected in leaves of wheat, barley, and the OK line. This indicates that aphids were able to feed on and transfer virus particles into leaves of the three species. However, after this incubation period, the virus particles were confined to the leaf tissue of the partial amphiploid and barley but were transported from the leaf to the stem below the leaf and to stems near the inoculated leaf of the wheat cultivars. At 12 and 18 days after inoculation, virus concentration in wheat was much higher than in barley or the partial amphiploid. For example, the number of stained bundles in the wheat cultivars increased from 7.8 to 17.3 in the leaf and from 23.0 to 37.7 in stems. In contrast, virus multiplication and movement in the partial amphiploid and barley was negligible. These observations indicate that the viral immunity in the partial amphiploid is due to a mechanism that reduces BYDV multiplication and inhibits systemic virus transport.

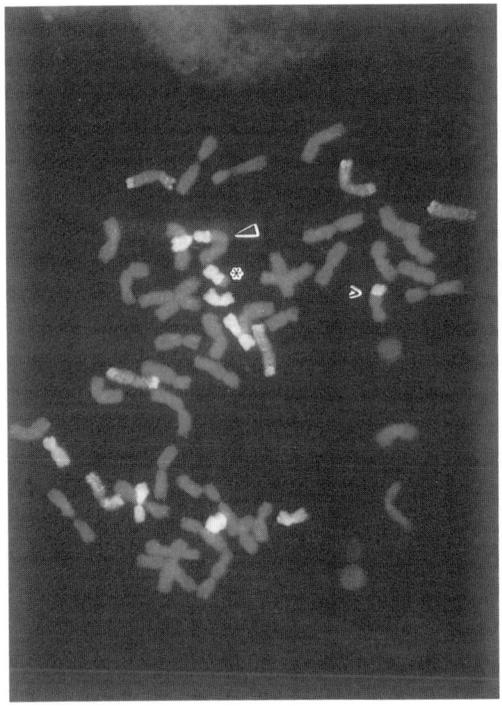

Figure 1. Somatic preparation of the partial amphiploid OT with 56 chromosomes, 42 from wheat and 14 alien chromosomes obtained from hybridization of wheat with *Thinopyrum intermedium*. The designation is partial amphiploid since the alien genome consists of chromosomes from both the St and J^s genomes of *Th. intermedium*.
The 42 wheat chromosomes are stained dark red; the 8 St genome chromosomes are light yellow; the 6 J^s genome chromosomes are pale red with yellow segments.
V -indicates a translocation involving entire wheat-St genome chromosomes; v -represents a wheat chromosome with a translocated terminal segment from a St genome chromosome; * -represents a St genome chromosome with a terminal translocated segment from a wheat chromosome.

The primary gene pool of wheat does not have any resistance to wheat streak mosaic virus and its vector the wheat curled mite (WCM). Two of the partial amphiploids shown in Table 1 had some degree of resistance to the above virus. Zhong 5 was highly resistant to WSMV and TAF46 was moderately resistant (Chen *et al.*, 1998). Both strains were susceptible to the WCM. Another partial amphiploid not included in Table 1, named 'Agrotana' was immune to both WSMV and WCM. Agrotana was obtained from crosses of wheat to *Th. ponticum* and contains 8 J^s and 8 J genome chromosomes (Chen *et al.*, 1998).

Table 2. Number of stained phloem bundles in shoots in tissue blots 16 days after inoculation in bread wheat, *Agrotricum* amphiploids and barley lines

Species and genotype	No. plants tested	Immune (No. plants with 0-1 stained bundles)	Resistant (No. plants with 2-10 stained bundles)	Susceptible (No. plants with 11-30 stained bundles)
Triticum aestivum				
cv. AC Pollet	20	0	0	20 [25]*
cv. AC Mimi	20	0	0	20 [21]
Agrotricum-amphiploid				
Zhong-4	10	0	10 [6]	0
TAF46	10	0	10 [7]	0
OK7211542	20	18 [0.1]	2 [2.5]	0
Hordeum vulgare				
line Q.B.172.15	4	4 [0.3]	0	0
line 80-81BQCB-10	4	4 [0.2]	0	0

* Figures within brackets represent mean number of stained phloem bundles.

In our earlier studies with GISH on *Th. intermedium* and *Th. ponticum* using St. genome as probe and E genome as a block we had established the genomic constitution of *Thinopyrum intermedium* as E_1 E_2 St (or J_1 J_2 St) and of *Th. ponticum* as EEE St St. The St. genome in *Thinopyrum intermedium* gave a clear, bright signal whereas one other genome of *Th. intermedium* and two genomes of *Th. ponticum* showed fluorescence near the centromeres of the two genomes. In other words, the *in situ* signals on these genomes were intermediate between the St. and J genomes. More recently Chen *et al.* (1998a) have proposed that those genomes with the St. signal in centromere regions, only be designated as J^s. Therefore the genomes of *Th. intermedium* are more correctly designated J J^s St. and that of *Th. ponticum* as JJJJ^s J^s. BYDV resistance coming from *Th. ponticum* therefore originates in the J^s genome and not St. as previously stated (Zhang *et al.*, 1996). The resistance from *Th. intermedium* also probably originates in the J^s genome.

As stated above, TAF46, one of the partial amphiploids derived from *Thinopyrum intermedium* is resistant to BYDV. Six of the possible 7 addition lines have previously been produced from that line. The BYDV resistance has been assigned to addition line 1 (L-1) or Ai #7 (Friebe *et al.*, 1992). This latter chromosome is somewhat longer than those in addition lines 4 and 7 that were derived from the same source. The chromosomes in addition lines 4 and 7, based on the bright, uniform FISH signal were definitely derived from the St genome. The additional chromosome in addition line 1, based on an St-type of signal in the centromere region was described as a complex translocated chromosome consisting of an St-origin centromere region flanked by J chromosome translocated arms (Wang and Zhang, 1996). More recently we have shown that the additional chromosome in line 1 was not translocated, but a typical J^s type of chromosome (Chen *et al.*, 1999). On that basis, we concluded that the BYDV resistance in TAF46 and addition line 1 originated from the J^s genome of *Th. intermedium*. Based on FISH analysis in a related study (Fedak *et al.*, 1999) it was shown that the partial amphiploid OK7211542 was derived from *Th. ponticum*. In that case the resistance was likely derived from the J^s genome. Of the partial amphiploids with BYDV resistance shown in Table 1 it would appear that the resistance in *Th. ponticum* derivatives is derived from the J^s genome whereas in *Th. intermedium* derivatives the resistance could originate from either the St or J^s or J genome but most likely the J^s genome.

This report shows that BYDV resistance in the partial amphiploids is far superior to that of wheat and is based on reduced virus multiplication and inhibition of virus transport through vascular tissue. The resistance in OK7211542 is superior to that of TAF46. Resistance in the latter has been studied in some detail (Brettel *et al.*, 1988) and transferred to wheat by recombination induced by callus cultures (Banks *et al.*, 1995). The superior resistance of OK7211542 is presently being manipulated in the same way (Chen *et al.*, 1998a).

References

Banks, P.M., P.J. Larkin, Bariana, H.S., Lagudah, E.S., Appels, R., Waterhouse, P.M., Brettell, R.I.S., Chen, X, Xu, H.J., Xin, X.Y., Qian, Y.T., Zhou, X.M., Cheng, Z.M. and Zhou, G.H. 1995. The use of cell culture for subchromosomal introgressions of barley yellow dwarf virus resistance from *Thinopyrum intermedium* to wheat. Genome. **38**: 395-405

Brettel, R.I.S., P.M. Banks, Y. Cauderon, X. Chen, Z.M. Chen, P.J. Larkin, and P.M. Waterhouse, 1988. A single wheatgrass chromosome reduces the concentration of barley yellow dwarf virus in wheat. Ann. Appl. Biol. **113**: 599-603

Cauderon, Y., B. Saigne, and M. Daige, 1973. The resistance to wheat rusts of *Agropyron intermedium* and its use in wheat improvement. In: Proc. 4th Int. Wheat Genet. Symp. University of Missouri, Columbia, Mo. pp.401-407

Chen, Q., J. Collin, R. Comeau, C.A. St.Pierre, and G. Fedak, 1997. Comparison of various sources of resistance to barley yellow dwarf virus in wheat - *Thinopyrum* amphiploids lines. Can. J. Plant Path. **19**: 414-417

Chen, Q., F. Ahmed, J. Collin, A. Comeau, G. Fedak, and C.A. St.Pierre, 1998a. Genomic constitution of a partial amphiploid OK7211542 used as a source of immunity to barley yellow dwarf virus for bread wheat. Plant Breed. **117**: 1-6

Chen, Q., R.L. Conner, F. Ahmad, A. Laroche, G. Fedak, and J.B. Thomas, 1998b. Molecular characterization of the genome composition of partial amphiploids derived from *Triticum aestivum* x *Thinopyrum ponticum* and *T. aestivum* x *Th. intermedium* as sources of resistance to wheat streak mosaic virus and its vector *Aceria tosichella*. Theor. Appl. Genet. **97**: 1-8

Chen, Q., R.L. Conner, A. Laroche, and J.B. Thomas, 1998c. Genome analysis of *Thinopyrum intermedium* and *Th. ponticum* using genomic *in situ* hybridization. Genome. **41**: 580-586

Chen, Q., R.L. Conner, A. Laroche, W. Ji, K.C. Armstrong, and G. Fedak, 1999. Genomic *in situ* hybridization analysis of *Thinopyrum* chromatin in a wheat *Th. intermedium* partial amphiploid and six derived chromosome addition lines. Genome (submitted)

Cisar, G., C.M. Brown, and H. Jedlinski, 1982. Effect of fall and spring infection and sources of tolerance of Barley Yellow Dwarf of winter wheat. Crop Sci. **22**: 474-478

Comeau, A. and J.P. Dubuc, 1977. Observations on the 1976 barley yellow dwarf epidemic in Eastern Canada. Can. Plant. Dis. Surv. **57**: 42-44

Comeau, A., K.M. Makouk, F. Ahmed and C.A. St. Pierre, 1994. Bread wheat x *Agrostricum* crosses as a source of immunity and resistance to the PAV strain of barley yellow dwarf luteovirus. Agronomie **2**: 153-160

Fedak, G., Q. Chen, R.L. Conner, A. Laroche, R. Petroski, and K.C. Armstrong, 1999. Characterization of wheat-*Thinopyrum* partial amphiploids by meiotic analysis and genomic *in situ* hybridization. Genome (submitted)

Forster, B.P., S.M. Reader, S.A. Forsyth, R.M.D. Koebner, T.E. Miller, M.D. Gale, and Y. Cauderon, 1987. An assessment of the homoeology of six *Agropyron intermedium* chromosomes added to wheat. Genet. Res. **50**: 91-97

Friebe, B., Y. Mukai, B.S. Gill, and Y. Cauderon, 1992. C-banding and *in situ* hybridization analysis of *Agropyron intermedium*, a partial amphiploid and six derived chromosome addition lines. Theor. Appl. Genet. **84**: 899-905

Gill, C.C., 1980. Assessment of losses on spring wheat naturrally infected with Barley Yellow Dwarf Virus. Plant Dis. **64**: 197-203

Li, Z.S., S. Rong, S. Chen, G.C. Zhong, and S.M. Mu, 1985. Wheat Wide Hybridization. Scientific Publishing Company, Beijing.

Sun, S.C. 1981. The approach and methods of breeding new varieties and new species from *Agrotriticum* hybrids. Acta Agron Sinica. **7**: 51-58. (Cited after Zhang *et al.* 1996)

Wang, R.R.C., R. von Bothmer, J. Dvorak, G. Fedak, I. Linde-Laursen, and M. Muramatsu, 1994. Genomic symbols in the Triticeae. In: Proceedings of the 2nd International Triticeae Symposium. Utah State University Press, Logan. pp.29-34.

Wang, R.R.C. and X.Y. Zhang, 1996. Characterization of the translocated chromosome using fluorescent *in situ* hybridization and random amplified polymorphic DNA on two *Triticum aestivum-Thinopyrum intermedium* translocation lines resistant to wheat streak mosaic or barley yellow dwarf virus. Chromosome Res. **4**: 583-587

Zhang, X.Y., A Koul, R. Petroski, T. Ouellet, G. Fedak, Y.S. Dong, and R.R.C. Wang, 1996. Molecular verification and characterization of BYDV-resistant germplasm derived from hybrids of wheat with *Thinopyrum ponicum* and *Th. intermedium*. Theor. Appl. Genet. **93**: 1033-1039.

In vitro techniques for genomic alteration in rice plants

T. Kinoshita and K. Mori
Faculty of Agriculture, Hokkaido University, Sapporo 060-8589, Japan

Key words: asymmetric somatic hybrids, *in-situ* hybridization, *Oryza* species, protoplast fusion, ribosomal DNA gene, rice blast resistance

Summary

Because of the explosive increase in world population, a sufficient food supply must be achieved by varietal improvement in the major cereal crops including rice. It is expected that new *in vitro* techniques in combination with conventional breeding methods may effectively raise the yield potential. On the other hand, there are many environmental problems to be solved world-wide such as, global warming, environmental pollution, ecological destruction, reduction in water supplies and so on. Therefore, it is necessary to rapidly develop new varieties for the future combining of higher yield potential with excellent grain quality, and resistance to both biotic and abiotic stresses for the promotion of sustainable agriculture. Although many efforts have been made to introduce useful traits from wild species to cultivated rice via hybridization, it is still difficult to overcome breeding barriers such as cross incompatibility and hybrid sterility and inviability in practical breeding. Now *in vitro* techniques are going to make it possible to use genetic manipulation and cell culture and fusion techniques to speed up the breeding process. For sustainable agriculture, it is important to utilize the useful genes from alien species. For this purpose, asymmetric protoplast fusions have already been used successfully to transfer disease resistance in *Brassica napus*. In this experiment, a high level of resistance to the rice blast disease was transferred from wild species through asymmetric fusions. It is also noted that manipulation of cytoplasmic genomes is possible through asymmetric fusions as shown in the induction of new cytoplasmic male sterility (CMS).

M. Maluszynski and K.J. Kasha (eds.),
Mutations, In Vitro and Molecular Techniques for Environmentally Sustainable Crop Improvement, 35–43.

Introduction

It has been proposed to urgently produce, or at least seek, an ideal crop genotype which would be suitable for the application of environmentally sustainable agriculture and in which advanced *in vitro* techniques are going to be efficiently used to speed up the breeding process. The recent development of high-density genetic linkage maps have made it possible to identify and unlock the genetic potential involved in wild species (Tanksley and McCouch, 1997). In rice, important genes for sustainable agriculture, such as the resistance to brown planthopper, bacterial leaf blight and grassy stunt virus, were introduced from some wild species (Brar and Khush, 1997). The use of *in vitro* techniques such as cell fusion and transformation can overcome the problems of crossability barriers and limited recombination to introduce the useful genes from distantly related species or genera. In this study, both symmetric and asymmetric protoplast fusions were carried out to produce somatic or cytoplasmic hybrids using rice cultivars and several *Oryza* species. The fusion products were evaluated for alteration of genomic constitution by using modern techniques. New approaches in chromosome research were also used efficiently for discrimination of different genomes and to detect loci of specific genes or gene families.

Effect of radiation dosages on asymmetric protoplast fusions

For induction of seed calli, cell fusion and plant regeneration in rice, the methods developed by (Kyozuda *et al.*, 1987) were followed. Protoplasts were isolated from the suspension culture of seed calli in several cultivated varieties and wild species, however green plant regeneration was only possible from protoplasts of cultivated rice. Plant regeneration was achieved on nurse cultures from rice or fused protoplasts. Prior to cell fusion, protoplasts isolated from the recipient cultivar 'Kitaake' were treated with the metabolic inhibitor 30 mM iodoacetamide (IOA) for 20 minutes at 4°C in 5% MES-KMC at pH 5.6 to prevent the cell divisions. Subsequently, an electrofusion apparatus (SSH-1, Shimazu) was used for the protoplast fusion.

The effect of gamma irradiation on suspension culture cells of *Oryza* species was studied. Doses of 0, 100, 200, 500, 800 and 1,000 Gy were applied at a fixed exposure rate of 15 Gy/minute. The highest regeneration rates from calli, following asymmetric fusion were achieved at doses of 800 Gy applied to the donor protoplasts. Higher doses of irradiation decreased the plating efficiency of fusion products. Thus an application of 800 Gy applied to donor protoplasts was considered an optimum dose with which to stimulate colony formation and plant regeneration.

Transfer of cytoplasmic male sterility (CMS) by donor-recipient method

The transfer of male sterile cytoplasm can be accomplished by an asymmetric fusion called the 'donor-recipient method' (Aviv *et al.*, 1984). Protoplasts isolated from gamma irradiated suspension cells of A-58CMS having the cytoplasm of [*cms-bo*] derived from 'Chinsurah boro II' were electrofused with those from Japanese cultivar, 'Fujiminori' after metabolic inhibition with iodoacetoamide.

In R_0 plants regenerated from the fused products, considerable variation in seed fertility ranging from completely sterile to nearly complete fertility was observed. It was estimated that completely sterile plants were caused by the combination between Fujiminori nuclear genome and [*cms-bo*] cytoplasm derived from A-58CMS. Further, it was noted that one of the completely sterile plants produced both male fertile and sterile clones when propagated asexually, suggesting the existence of a heteroplasmic state in the original R_0 plant (Izhar *et al.*, 1983).

With the exception of two plants, a stable transmission of CMS was demonstrated in most of the R_0 plants. Later, it was found that both exceptional plants consisted of a single copy of the *atp6* gene in the mitochondrial genome that is equivalent with those of Fujiminori. In the male fertile R_0 plants, the crossing experiments showed there was no alteration of the nuclear genotype (rf_1rf_1) from Fujiminori. Therefore, it was implied that the sorting out of sterile elements such as the first heteroplasmic state produced various degrees of fertility restoration in R_0 plants (Izhar *et al.*, 1983).

Kadowaki *et al.* (1990) demonstrated that the extra *atp6* gene in mtDNA of [*cms-bo*] cytoplasm plays an important role in CMS expression, and the chimeric *atp6* gene was named as *urf-rmc*. In the male sterile plants, two copies of *atp6* equivalent with A-58CMS were included. However, two plants from male sterile R_0 plants contained an extra signal of *atp6* in addition to the two confirmed to be in A-58CMS. There is a possibility that a new constitution of mtDNA was caused by a recombination between the parental mtDNAs in these plants. On the other hand, male fertile R_0 plants consisted mostly of the pattern equivalent to that in Fujiminori. Thus the nature of *atp6* gene corresponded fairly well with the expression of CMS in male sterile R_0 plants. However, it was newly found that the 10.4 kb-Xho I fragment containing the *rps1* gene in the mtDNA was also closely related to CMS expression together with the chimeric *atp6* gene. Transfer of cytoplasmic male sterility from A-58CMS was also achieved in the promising strain, `Kochihibiki' and it is possible to use this method for the practical breeding of hybrid rice.

Transformants produced by asymmetric fusion

Protoplasts isolated from W124 (*O. rufipogon*) irradiated with gamma rays were electrofused with those from rice cultivar Kitaake. Several plants were regenerated, and the transfer of cytoplasm from W124 to Kitaake was confirmed by Southern hybridization probed with B3, one of the plasmid-like DNAs involved in the mitochondrial genome of W124.

Although most of the R_0 plants showed Kitaake plant type irrespective of the ploidy (2*x* or 4*x*), one R_0 plant, RU-75 indicated a spreading plant type, purple apiculus color and red seed coat which resembled those of W124, while awn length and spikelet shape were intermediate between both parents. According to RFLP analysis of RU-75 using the combinations between 86 RFLP probes and 7 restriction enzymes, RU-75 showed banding patterns of Kitaake type for 79 markers. In addition, it had those of W124 type for 2 markers and a unique type different from the both parental types for 5 markers, and all of the markers showed homozygous condition. Furthermore all characters were transferred from W124 bred true in the R_1 and later generations.

Thus, the transfer of genes from *O. rufipogon* to (AA, 2n=24) the cultivated rice was achieved in RU-75, although the genetic mechanism was unknown. There is also a possibility that somaclonal variation or homozygous mutation (Oono, 1985) occurred in several parts of chromosomes of the somatic hybrid.

Genetic construction of the plants regenerated from the symmetric or the asymmetric fusion

In the combination between Kitaake (recipient) and *O. punctata* (donor), R_0 plants from the symmetric fusion showed a wide range of chromosome numbers as shown in Table 1.

Table 1. Chromosomal variation in R_0 plants produced by the symmetric fusion

Chromosome numbers (2n)	24	48	60~71	72	>84	Total
Frequency	1	2	5	6	2	16

Mixoploidy was frequently observed especially in aneuploid R_0 plants. It is expected that a read hybrid between rice and *O. puctata* consists of 2n=72. According to the molecular cytogenetical observation, six R_0 hexaploids consisted of AA genome from rice and BBCC from *O. punctata*. Morphological features of the somatic hybrids over tetraploids indicated an intermediate type between those of both parents, while diploid and tetraploid R_0 plants showed Kitaake plant type. However the consitution of 45S rDNA of somatic hybrid showed a variation even in a diploid R_0 plant. Because of complete sterility in the somatic hybrids over diploid and tetraploid, no seeds were obtained for R_1 progeny tests. In the back fusion experiment, the progenies from a hexaploid hybrid indicated a chromosomal variation ranging from 61 to 72 after the symmetric fusion with Kitaake, and all plants were still completely sterile. Restoration of fertility was possible in the progenies when the hexaploid hybrid was back fused by asymmetric fusion with Kitaake.

Progeny tests were carried out using one tetraploid and nine diploid R_0 plants showing Kitaake plant type. Although high fertility was restored in some R_1 or R_2 lines, the segregation of seed fertility still continued in most of the lines. Several off-type plants such as extremely dwarf and elongated culm structure also segregated in several R_1 or R_2 lines.

In the asymmetric fusion from the same combination, R_0 plants consisted of 73 diploids, 9 tetraploids and one heptaploid ($2n$ = 84). In RFLP analyses using 12 chromosome markers and one DNA fragment from ctDNA, most of the diploid and tetraploid plants showed the Kitaake type except for minor variations. In contrast with this, diversity of mtDNA due to segregation and recombination was detected in six of 64 R_0 plants by RFLP analyses using the probes, *atp6* and *coxII* from mtDNA. An effect of radiation dosages shown in tomato (Melzer and O'Connel, 1992) was applicable at 80 krads in rice. Thus it was demonstrated that cytoplasmic hybrids are involved among R_0 plants as seen in other cases (Walters and Earle, 1993; Wolters *et al.*, 1995). Enough seeds for R_1 progeny tests were obtained from most of the R_0 plants derived from the asymmetric fusion.

Rice blast resistance in R_1 lines

By using the R_1 lines obtained from six kinds of asymmetric fusions of Kitaake with other *Oryza* species, the resistance in each line was evaluated by a field inoculation test that is widely used for the selection of resistant lines in rice breeding. In the frequency distributions of R_1 lines from the asymmetric combinations, it is noted that a wide range of variation over those of

Kitaake (m and m-r) was obtained in all combinations as shown in Table 2. Particular R_1 lines showed highly resistant lines to rice blast fungus that appeared in the two combinations with *O. punctata* and *O. grandiglumis*. Although further confirmation is needed for the nature of the alteration of genetic constitution in the resistant plants, there is a possibility that cytoplasmic effects might be responsible for enhancing the grade of resistance.

Table 2. Number of rice blast resistance grades in R_1 lines derived from asymmetric fusions between rice and *Oryza* species

Combination	Grade of resistance*						
Donor Recipient	S	s	m	m-r	r	R	Total
O. rufigogon+Kitaake	2		10	9			21
O. minuta+Kitaake		1	6	4			11
O. glaberrima+Kitaake		2	2	5	2		11
O. grandiglumis+Kitaake		2	7	8	20	18	55
O. eichengeri+Kitaake		1	1				2
O. punctata+Kitaake	1	7	4	13	19		44
Total	3	13	30	39	41	18	144

*S (highly susceptible), s (susceptible), m and m-r (moderately resistant), r (resistant) and R (highly resistant)

Induction of novel CMS type

In the former experiment, CMS was successfully transferred into cultivated rice from the donor variety, A-58CMS through asymmetric fusions. There is also a possibility that a new type of CMS can be induced in new combinations between nuclear and cytoplasmic genomes. In this experiment, seed fertility was examined in R_0 plants from the four combinations of asymmetric fusions between Kitaake and *Oryza* species. Except for *O. glaberrima*, cytoplasms donated from *O. rufipogon*, *O. minuta* and *O. punctata* were effective in the generation of completely sterile R_0 plants. Although crossing experiments are needed to define the determination of sterility, cytoplasm from *O. rufipigon* was most efficient in inducing a novel CMS type (Table 3).

Application of molecular cytogenetics

Diagnosis by GISH

The application of an *in situ* hybridization method (ISH), has made it possible to identify different genomes in plants. This method, called genomic *in situ* hybridization (GISH), is extensively used for identification of chromosomes belonging to different genomes in the progeny of somatic hybrids between tomato and wild eggplant (Escalante *et al.*, 1998). In our study, multi-color genomic *in situ* hybridization, McGISH, using three different fluorescent colors to discriminate the three different genomes, was firstly applied for the somatic hybrids produced between Kitaake and *O. punctata*, and A, B and C genomes were clearly discriminated.

Table 3. Male sterile R_0 plants induced by asymmetric fusions between rice and *Oryza* species

Combination		Seed fertility				
Donor	Recipient	Sterile	Low	Middle	High	Total
O. punctata+Kitaake		2(1)	4(2)	9(2)	21	36
O. glaberrima+Kitaake		0	3	7	10	20
O. minuta+Kitaake		2	1	2	1	6
O. rufipogon+Kitaake		31(1)	5	5(1)	5	46

Parentheses mean tetraploids

Two somatic hybrid plants having chromosome numbers approximately 72 were produced from symetric fusion between rice Kitaake and *O. punctata*, and used as materials. PU-2 consisted of a pure hexaploid ($2n$ = 72) and PU-289 was mixoploid, having somatic chromosome numbers between 65 and 72. In PU-2 plant, seventy-two chromosomes were visualized in most of the cells by counter-staining with DAPI. Twenty-four red chromosomes belonging to the A genome and 24 green chromosomes belonging to the C genome were detected individually. B genome was identified by counter-staining applied with DAPI as blue signal. Finally, the 24 chromosomes derived from each of the A, B and C genomes were completely identified. In the nuclear region, each genome seemed to occupy its own domain around the nucleolus, which was unstained as if it were a small cavity. Sixty-five to seventy-two chromosomes of PU-289 were counted after counter-staining. Based on Multicolor GISH (McGISH), the chromosome number of 24 expected from A genome was always maintained in every nuclear plate. On the other hand, a loss of the chromosomes from the B and C genomes was frequently observed. For example, 22 and 23 chromosomes belonging to the B and C genomes, respectively, were detected in a total of 69 chromosomes. Two cases of insertion of the B genome chromosomal fragments to the C genome chromosome and of A genome chromosomal fragments into B genome chromosome were detected. These insertions and translocations were also observed frequently. In the nuclei, each genome occupied its own domain around the nucleolus as in the case of PU-2.

Jacobsen *et al.* (1995) employed the GISH method for the identification of alien chromosomes in the backcross progeny of fusion hybrids between potato and tomato, and confirmed the usefulness of this method for plant breeding programs. The McGISH method shows considerable possibilities for the characterization of various somatic hybrids and chromosomal diagnosis for the progenies derived from both somatic and sexual hybridization.

Identification of rice D-genome chromosomes

The 24 D genome chromosomes were identified using GISH, which uses total genomic DNA as the probe among the 48 chromosomes of *O. latifolia*, comprised of the C and D genomes. The B genome chromosomes were also discriminated from the C genome chromosomes in *O. minuta* (BBCC) by GISH. A comparison of the fluorescence intensity between the C and D genomes within *O. latifolia* (CCDD), and between the B and C genomes within *O. minuta*, indicated that the nucleotide-sequence homology between the B and C genomes is less than that between the C and D genomes. Thus, the phylogenetic relationships among B, C and D genomes were estimated.

Physical mapping of ribosomal RNA genes (rDNAs)

Rice chromosomes with a large (17S-5.8S-25S subunits) ribosomal RNA gene (45S rDNA) have been identified as the satellited chromosomes by their karyotype. They are also recognized as the chromosomes with a nucleolar organizing region (NOR chromosomes). On the other hand, the 5S rDNA genes are generally not incorporated in such repeating units, but are located at numerous chromosomal sites.

Development of fluorescence *in situ* hybridization (FISH) improved the efficiency and sensitivity of detection. Fukui *et al.* (1994) detected one 45S rDNA locus in *O. sativa* (*japonica*), *O. rufipogon* and *O. brachyantha*. According to them, two 45S rDNA loci were found in *O. sativa* (*indica* and *javanica*), *O. rufipogon* and *O. australiensis*. Three loci were revealed in *O. officinalis*. 5S rDNA has also been mapped by FISH on chromosome 9 of *indica* rice (Song and Gustafson, 1993) and on chromosome 11 of *japonica* rice (Kamisugi *et al.*, 1994). The 5S rDNA locus was detected on the proximal region of the short arm in *japonica* rice. Further research by FISH on localization of 5S rDNA using several *Oryza* species revealed that most of the species examined had only one locus on chromosome 11 except for *O. australiensis* and *O. brachyantha* both of which had the one locus on the short arm of chromosome 7. Ohmido and Fukui (1995) developed a multicolor FISH (McFISH) method using 45S and 5S rDNA probes simultaneously in the African cultivated rice, *O. glaberrima* (A^gA^g) and detected identical localization of both the rDNAs on chromosome 11 as in *O. sativa* ssp. *japonica*.

According to the present observation, the 45S rDNA loci were detected on chromosome 4, 9 and 11 in *O. officinalis* and *O. eichingeri*. The loci on chromosomes 4 and 9 were localized at the distal end of short arms and the signal intensity of chromosome 4 was stronger than that of chromosome 9. Further, the third 45S rDNA locus was the weakest among the three rDNA loci and identified at the telomeric region neighboring the 5S rDNA locus on the short arm of chromosome 11. In bothe *O. australiensis* and *O. brachyantha*, loci of 45S and 5S rDNA were confirmed on chromosome 7 and 9 respectively. McFISH simultaneously revealed the number and position of both the 45S and 5S rDNA loci mentioned above using Dig-5S rDNA and Bio-45S respectively as probes. From these results (Table 4), the variation of rDNA loci among *Oryza* species has evolutionary significance indicating, that structural change or transposition might be involved during species differentiation.

Table 4. Diversity of rDNA loci in *Oryza* spcies

Genome	Species	Distribution	Chromosome				
			4	7	9	10	11
AA	*O. sativa* ssp. *japonica*	Temperate region			45S		5S
AA	*O. sativa* ssp. *indica*	Subtropics and tropics			45S	45S	5S
BB	*O. punctata*	Africa			45S		5S
CC	*O. officinalis*	Tropical Asia	45S		45S		5S, 45S
CC	*O. eichingeri*	Central Africa	45S		45S		5S, 45S
EE	*O. australiensis*	Tropical Australia		5S	45S		
FF	*O. brachyantha*	Africa		5S	45S		

Prospects

Applications of molecular cytogenetics including FISH and GISH techniques are important for the evaluation of somatic hybrids and their progenies. Chromosome engineering can also be utilized more efficiently for practical breeding by applying these methods. Because the new loci of both 45S and 5S rDNA were detected among *Oryza* species, diversification of this important gene or gene clusters may be helpful in elucidating the phylogenetic differentiation among the species of *Oryza*. Visual observation of gene or gene families on the chromosomes will contribute to improve breeding efficiency, and can be utilized in many ways for molecular and cell breeding.

References

Aviv, D., P. Arzee-Gonen, S. Bleichman, and E. Galun, 1984. Novel alloplasmic *Nicotiana* plants by "donor-recipient" protoplast fusion: cybrids having *N. sylvestris* nuclear genomes and chondriomes from alien species. Mol. Gen. Genet. **196**: 244-253

Brar, D.S. and G.S. Khush, 1997. Alien introgression in rice. Plant Mol. Biol. **35**: 35-47

Escalante, A., S. Imanishi, M. Hossain, N. Ohmido, and K. Fukui, 1998. RFLP analysis and genomic *in situ* hybridization (GISH) in somatic hybrids and their progeny between *Lycopersicon esculentum* and *Solanum lycopersicoides*. Plant Mol. Biol. **35**: 35-47

Fukui, K., N. Ohmido, and G.S. Khush, 1994. Variability in rDNA loci in the genus *Oryza* detected through fluorescence *in situ* hybridization. Theor. Appl. Genet. **87**: 893-899

Izhar, S., M. Schlicter, and D Swartzberg, 1983. Sorting out of cytoplasmic elements in somatic hybrids of petunia and prevalence of the heteroplasmon through several meiotic cycles. Mol. Gen. Genet. **190**: 468-474

Jacobsen, E., H.J. de Jong, S.A. Kamstra, P.M.M.M. Van Den Berg, and M.S. Ramanna, 1995. Genomic *in situ* hybridization (GISH) and RFLP analysis for the identification of alien chromosomes in the backcross progeny of potato (+) tomato fusion hybrids. Heredity. **74**: 250-257

Kadowaki, K.T., T. Suzuki, and S. Azama, 1990. A chimeric gene containing the 5 portion of *atp6* is associated with cytoplasmic male-sterility of rice. Mol. Gen. Genet. **224**: 10-16

Kamisugi, Y., S. Nakayama, R. Nakajima, H. Ohtsubo, E. Ohtsubo, and K. Fukui, 1994. Physical mapping of the 5S ribosomal RNA genes on rice chromosome 11. Mol. Gen. Genet. **245**: 133-138

Kyozuka, J., H. Hayashi, and K. Shimamoto, 1987. High frequency plant regeneration from rice protoplasts by novel nurse culture method. Mol. Gen. Genet. **206**: 408-413

Melzer, J.M. and M.A. O'Connell, 1992. Effect of radiation dose on the production of and the extent of asymmetry in tomato asymmetric somatic hybrids. Theor. Appl. Genet. **83**: 337-344

Ohmido, N. and K. Fukui, 1995. Cytological studies of African cultivated rice, *Oryza glaberrima*. Theor. Appl. Genet. **91**: 212-217

Oono, K., 1985. Putative homozygous mutations in regenerated plants of rice. Mol. Gen. Genet. **198**: 377-384

Song, Y.C. and J.P. Gustafson, 1993. Physical mapping of the 5S rDNA gene complex in rice (*Oryza sativa*). Genome. **36**: 658-661

Tanksley, S.D. and S.R. McCouch, 1997. Seed banks and molecular maps: unlocking genetic potential from the wild. Science. **277**: 1063-1066

Walters, T.W. and E.D. Earle, 1993. Organellar segregation, rearrangement and recombination in protoplast fusion-derived *Brassica oleracea* calli. Theor. Appl. Genet. **85**: 761-769

Wolters, A.M.A., H.C.H. Schoenmakers, and M. Koornneef, 1995. Chloroplast and mitochondrial DNA composition of triploid and tetraploid somatic hybrids between *Lycopersicon esculentum* and *Solanum tuberosum*. Theor. Appl. Genet. **90**: 285-293.

An improved *in vitro* technique for isolated microspore culture of barley

K.J. Kasha, E. Simion, R. Oro, Q.A. Yao[1], T.C. Hu[2] and A.R. Carlson[3]
Dept. of Plant Agriculture, Biotechnology Divison, University of Guelph, Guelph, ON Canada, N1G 2W1
[1] *Present address: Celera Genomics, 45 West Gude Dr., Rockville, MD 20850, U.S.A.*
[2] *Present address: Monsanto Life Sciences, 700 Chesterfield Pkwy N., St. Louis, MO 63198, U.S.A.*
[3] *Present address: Dept. of Agronomy, University of Wisconsin, Madison WI 53706, U.S.A.*

Key words: barley, cell culture, haploid, induction, microspore, regeneration

Summary

A detailed procedure for isolated microspore culture of barley is presented along with examples of response across genotypes. Over 30 genotypes, including winter and spring growth habit and 2-row and 6-row genotypes, have shown an essentially genotype independent response, averaging about 10,000 embryos per 5 cm petri culture plate. The regeneration frequencies, checked on samples of 500 embryos per plate ranged from 36 to 97% with most genotypes being in the range of 70 to 90%. About 70 to 80% of the plants regenerated have been completely fertile doubled haploids, thus eliminating the need to double the chromosome number of plants. Many little details are critical to success of the microspore procedure and while it saves much time compared to anther culture, greater attention to details and cleanliness is essential.

M. Maluszynski and K.J. Kasha (eds.),
Mutations, In Vitro and Molecular Techniques for Environmentally Sustainable Crop Improvement, 45–54.

Introduction

The production of haploid plants (plants with the gametic chromosome number (*n*)) or doubled haploid plants (plants or cells that were haploid but have their chromosome number doubled to the normal somatic number (*2n*) via duplication) have long been recognized and studied for their application in plant breeding and other research. Many systems for haploid production have been developed (Kasha, 1974; Davies and Hopwood, 1980; Jain *et al.*, 1996). However, the system for inducing the thousands of microspores per floret to divide and produce embryos has been recognized as the one with the greatest potential to produce large numbers of haploids and doubled haploids. Methods using microspores have been developed where some genotypes within a species can be induced to produce large numbers of haploids. The limitation in applications has been the failure to produce large numbers across most or all genotypes of the species (Kasha *et al.*, 1995; Sopory and Munshi, 1996). Research has led to the description of three stages that may independently control haploid production from microspores, namely; the induction of embryogenesis; the regeneration of embryo-like structures; and their development into green plants (genotypes of some monocotyledonous species such as cereals often produce albino plants).

It has also been suggested that the haploid single-celled microspores would be an attractive target for mutation, selection or transformation (Kasha *et al.*, 1990; Maluszynski *et al.*, 1996); particularly if performed at the G1 phase of the nuclear cycle. The latter could lead to the production of plants genetically homozygous, including the transgenes. Transformation of isolated barley microspores has been achieved via biolistic bombardment (Jähne *et al.*, 1994; Yao *et al.*, 1997) of the highly inducible barley genotype 'Igri.' Jähne *et al.* (1994) used a cold pretreatment that resulted in plants homozygous for the transgene whereas, Yao *et al.* (1997) used mannitol pretreatment which resulted in heterozygous transgenics. Both used herbicide selection for the transgenic microspores. We have also used antibiotic selection and visual selection with green flourescent protein (GFP) genes to obtain transgenic plants from bombarded microspores.

After a number of years of effort, we have developed an isolated microspore culture system for barley using defined medium, which can produce large numbers of embryos from a wide range of genotypes. As described in earlier reviews (Jähne and Lörz, 1995; Dunwell, 1996) many labs have discovered key elements to the system, such as donor plant growth (Hunter, 1988), sugar type (Hunter, 1988; Scott and Lyne, 1994), nitrogen source and balance (Olsen, 1987; Mordhorst and Lörz, 1993), high glutamine (Olsen 1987) hormones (Ziauddin *et al.*, 1992), density and osmolality (Hoekstra *et al.*, 1993) and pretreatments (Roberts-Oehlschlager and Dunwell, 1990). The process has been streamlined with improved pretreatment and isolation procedures so that it is much more efficient and labour saving than anther culture procedures. In this paper we will describe these procedures in detail.

Procedures

Donor plants growth conditions

Plants are grown in a pest free environment, best in controlled conditions using a combination of cool-white Gro-Lux or incandescent bulbs to provide a 350-400 µE m^2s at pot level, relative humidity of 60-80% and temperatures

for: winter barley are 15°C during 16-17 hr light and 12°C dark/day
spring barley are 18°C during 16-17 hr light and 15°C dark/day

Plants are fertilized weekly and watered carefully to avoid water or drought stress. Stressed plants, through watering or pest infestation, often fail in microspore culture or respond inconsistently. Spraying with pesticides near the time of anther collection will often eliminate microspore response.

Harvest of spikes and pretreatment

Spikes are collected when microspores are at the mid- to late-uninucleate stages. Good descriptions and illustrations of stages have been published (Wheatley *et al.*, 1986; Hu, 1997) and one of the clearest staging features is the location of the nucleus relative to the microspore pore (Figure 1). In the early uninucleate stage, the nucleus is near the pore and by late uninucleate, it is opposite the pore. Staining the microspores of an anther in acteo-carmine works well in barley but Alexander's stain is better for wheat and maize. A floret from the central part of the spike is checked for staging. Tillers, close to the right stage based upon their developmental similarity to those staged, are cut and placed in H_2O and left at 4°C until spikes are removed from the sheath. This storage may be as long as 3 or 4 days while collecting sufficient spikes for an experiment, but this time should be minimised as some microspore stage progression can occur. Upon removal from the sheath, the spikes may be selected further based upon their length and floret appearance. In a flow bench, cut tillers are sprayed with 70% EtOH, spikes are removed, the awns pulled off and the spikes are placed in large petri plates (e.g. 15 x 100 mm) with 15 ml of ice-cold 0.3 M mannitol (enough to partially cover spikes arranged in a single layer).

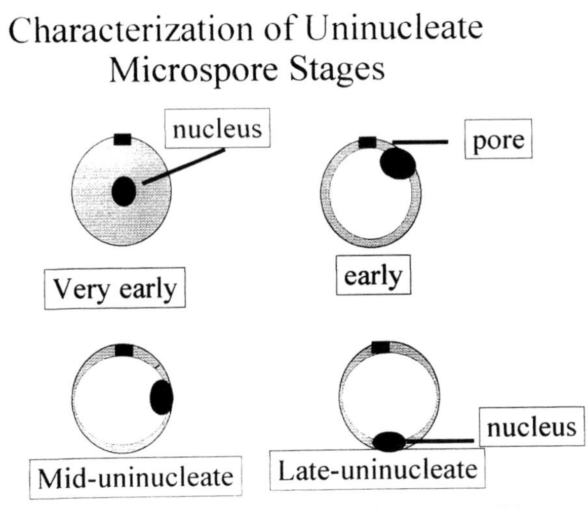

Figure 1. An illustration of the microspore stages based on the position of the nucleus relativ to the pore. Vacuole size is often considered as well.

Plates are sealed with a film (Parafilm, or a similar film) and placed in the dark at 4°C fo 3 to 5 days. It may be necessary to sterilize the spikes with 10-15% bleach (0.5 to 1.0%

chlorine) prior to mannitol pretreatment if awns or spikes have emerged from the leaf sheath or if contamination is a problem. Do no treat the isolated anthers with bleach. After bleach, rinse the florets or spikes well and allow any moisture to evaporate before adding the mannitol for pretreatment.

Pretreatment can be achieved by placing the spikes with 0.5 ml of sterile water in a petri dish at 4°C for 3 to 4 weeks. We have also found that the combination of cold (4° C) plus mannitol (0.3 M) is very effective for spike pretreatment and appears to result in a higher proportion of green regenerants than other pretreatments.

For anthers pretreatment, leaving them in the 0.3 M mannitol at room temperature (or 25°C) for 3 or 4 days and vortexing for 5 minutes at the time when the anthers are ready to shed has been effective (Kasha *et al.*, 1995). Do not treat the isolated anthers with bleach.

Microspore isolation

In a flow bench, using all sterilized (autoclaved) glassware and tools, pick up the pretreated spikes with forceps and use scissors to cut 2-3 cm segments into a chilled Waring Blender stainless steel mini sample container. Eight to ten 2-rowed spikes or four 6-rowed spikes provide sufficient microspores (500,000) for one culture plate. Add sufficient ice-cold 0.3 M mannitol to provide a slurry during blending (usually a ½ cup or enough to cover the spike segments). Blend at low speed for 6-10 seconds. Blending speed and time are critical and if microspore viability is low, this could be an area to experiment with. Filter the resulting slurry through cheesecloth or a coarse nylon membrane. Collect the filtrate and filter it through a 100 μ nylon mesh into a graduated 15 to 40 ml centrifuge tube. Spin 5 minutes at 900 rpm to provide a pellet of microspores. About 0.15 ml of pellet is approximately 500,000 barley microspores. Decant and resuspend the microspores in cold 0.3 M mannitol for washing and recentrifugation. Usually two or three washes are beneficial to microspore viability but we prefer to use as few as possible and stop when the wash supernatant does not show a green tinge.

Decant and resuspend microspores in 2 ml liquid FHG culture media. If the microspores are plasmolized, they will not be worth culturing. If the viability is below 50%, we would recommend removing the dead microspores by placing them on a 20% maltose density gradient and centrifugation. In 1995, we produced a 10 minute video of isolated barley microspore culture which is available from Dr. M. Maluszynski at IAEA/FAO, Vienna. This video illustrates the vortex isolation procedure from pretreated anthers, an alternate procedure to blending that can be used when the number of spikes available is very limited.

Microspore culture media

A great deal of research has gone into culture media in many labs and culture media is usually different in each lab. In our lab the FHG media of Hunter (1987), modified by using PAA, provides the best induction response and its ingredients are shown in Table 1. Key features are high glutamine, low NH_4NO_3, maltose as the sugar and phenyl acetic acid (PAA) as the auxin. The pH is adjusted to 5.8 and to solidify the medium, 3 gm of Phytagel are added to the liquid formula. Gelrite is also suitable for solidification but during selection much higher levels of antibiotic or herbicide are required to get effective selection, compared to Phytagel.

Once the embryos reach the 1 to 2 mm size, usually after four weeks, they are transferred to a differentiation media for a week or two until shoots and roots are developed. Then they are transferred into MS medium for regeneration (Table 1). The differentiation medium is a modified MS medium which contains 30 gm of maltose, 250 mg of inositol, 1000 mg of casein, 690 mg of proline, as well as the normal MS levels of thiamine, pyridoxine, nicotinic acid and BAP, but no auxin. It is solidified with Phytagel at 3 gm/l after adjusting the pH to 5.8. The smaller structures are left on the induction plate and a drop or two of freshly prepared liquid FHG medium are added to the plate. The solidified FHG medium without auxin could also be used for differentiation. The modified MS medium was developed for differentiation when we were using selection for Basta herbicide resistence in culture because the high glutamine reduced the effectiveness of selection.

Table 1. Culture media used for induction and regeneration of isolated barley microspores

Media components	FHG media induction (mg/l)	MS media regeneration (mg/l}
Macro salts		
KNO_3	1900	1900
NH_4NO_3	165	1650
KH_2PO_4	170	170
$MgSO_4 \times 7H_2O$	370	370
$CaCl_2 \times 2H_2O$	440	440
Micro salts		
$FeNa_2 \times EDTA$	40	37.3
$MnSO_4 \times 5H_2O$	22.3	22.3
H_3BO_3	6.2	6.3
$ZnSO_4 \times 7H_2O$	8.6	8.6
$CoCl_2 \times 6H_2O$	0.025	0.025
$CuSO_4 \times 5H_2O$	0.025	0.025
$NaMoO_4 \times 2H_2O$	0.25	0.25
KI	-	0.83
Other components:		
Glutamine	750	146
Inositol	100	100
Thiamine	0.4	0.1
Maltose	60.0 gm	-
Phenylacetic acid	10. mg	-
BAP	1.0	1.0
Pyridoxine	-	0.5
Nicotinic acid	-	0.5
Phytagel	-	3.0 gm
Sucrose	-	30.0 gm

Microspore plating and culture

In a vacuum filter flask, place two sterilized No. 2 Whatman filter papers, rinse them with liquid FHG medium, 0.3 M mannitol, or 10-20 ml sterile distilled H_2O. Rinsing the filters is an important step that may influence microspore response. Under low vacuum, pipet in drops 1 to 2 ml of FHG with microspores (500-600,000) on to the filter paper. Transfer the upper filter paper with microspores to solidified FHG media and seal plates with Parafilm or similar material. Some researchers find that using more than two wraps of Parafilm will cause culture response failure, presumably due to lack of gas exchange. Another method is to place small covered culture plates in a large petri plate along with a small open plate containing drops of distilled H_2O and then seal only the large plate. Culture in the dark at 25 to 28°C for 3-4 weeks, add drops of liquid FHG as paper dries out. Microspores may also be cultured in liquid FHG or on solidified FHG, however, we obtain better regeneration when using the filter paper or a membrane on top of solid media.

Regeneration

As mentioned previously, the 1-2 mm diameter embryos are transferred to fresh modified MS solid medium for continued development. Once shoots appear, they can be transferred to MS regeneration media (Table 1). Continue to culture in the dark for 3-4 days and then under low light intensity for 8 h/day at 22°C. Transfer small plantlets (3-4 cm tall) onto MS regeneration media without hormones in large vials or vessels to allow plants to grow. Once they reach 3 to 4 leaves, place the cultures under strong light for conditioning and after 3 or 4 days, transfer them into soil or potting media in pots.

Results

More than 30 different barley genotypes have been examined with the above described procedures and, in general, between 5,000 to 15,000 are produced per plate. Table 2 gives some actual counts of embryos based on an average of three replications (plates) but the enormity of the task of embryo counting limited the numbers of genotypes counted. Instead, response was judged on the appearance of plates after two weeks and again after 4 to 5 weeks, the latter bein illustrated for Igri and Trinity in Figure 2. Random samples of 500 embryos were taken from the counted plates and placed on regeneration media to get an estimate of the embryo regeneration ability. These have ranged from 36 to 97% (Table 2). Genotypes Igri, GBC777 and GBC778 are winter growth habit while the remaining lines in Table 2 are spring habit genotypes. 'Harrington', 'Manley' and 'Oxbow' are the predominantly grown 2-row malting barleys in Canada at present. Igri and 'Lina' are of European origin. Lines GBC 777 and 778 are 6-row barley.

Three different pretreatment methods were compared with the genotype Igri and no apparent differences were observed relative to embryo induction or regeneration (Table 2). Lina normally responds well in microspore culture following a mannitol pretreatment, but in this study it was low using the combination of cold plus mannitol for 7 days. Thus, embryo formation was also counted in Lina when a 21 days cold pretreatment had been used and this response was more typical of Lina (Table 2). The regeneration response from the two pretreatment methods on Lina was similar. The frequency of fertile doubled haploids among the Lina progeny following the two pretreatments was not different, namely 79% after the combination treatment and 83% after the 21 day cold pretreatment.

The response from Harrington was also low in both embryo formation as well as perentage regeneration. This genotype was not available in another experiment at that time, but more recently, the response has been observed to be consistently much better. Oxbow gave an interesting response. At the shorter times of pretreatment, it gave good response but the induced structures mostly became calli. However, when the longer pretreatment was used as shown in Table 2, the response was mostly embryogenic and resulted in fairly good regeneration. This response appears to be typical of Oxbow and leads us to speculate that it may have a higher endogenous auxin level but this has not been checked. The 6-row winter habit genotype GBC777 was unusual in that it produced mostly albino plants (about 5% being green). Even at that low rate, one could still expect to obtain about 500 green plants per plate if attempts were made to regenerate all the embryos. This would be labourious but one could transfer all the embryos to a suspension culture in regeneration media and low light to recover the green germinating embryos. The effect of the length of the cold plus mannitol pretreatment was examined for the genotype Igri (Table3) and was found not critical to response. However, it did appear to have an effect on Oxbow as previously mentioned (Table 2).

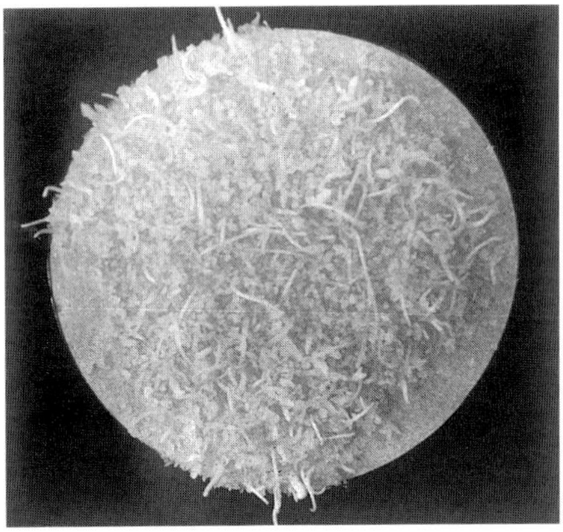

A

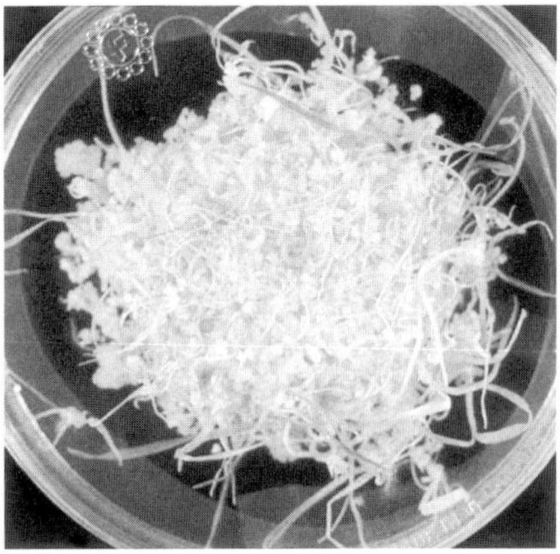

B

Figure 2. Embryo development in isolated microspore culture 4 to 6 weeks after culture initiation. A. cv. Igri, ca. 4-5 weeks. B. cv. Trinity, ca. 6 weeks.

Table 2. Embryo induction and regeneration following pretreatment of various genotypes in barley

Genotype	Pretreatment	Embryoids* (No.)	Embryoid regeneration (No.)
Igri	28 d cold, 4°C	13,136	85
	4 d 0.3 M man. 4°C	12,380	97
	4 d 0.3 M man. 25°C	13,504	88
GBC 782	3 d 0.3 M man. 4°C	14,496	n.a.
GBC 783	3 d 0.3 M man. 4°C	11,360	n.a.
Harrington	3 d 0.3 M man. 4°C	5,533	36
Lina	21 d cold 4°C	15,377	77
	7 d 0.3 M man. 4°C	8,921	81
Oxbow	21 d 0.3 M man. 4°C	10,810	58
Manley	4 d 0.3 M man. 4°C	11,360	74
GBC 777	4 d 0.3 M man. 4°C	14,133	89
GBC 778	4 d 0.3 M man. 4°C	7,667	84

*Embryo numbers are the average from 3 plates while percentage of regeneration is based on shoots formed on 500 embryos per plate; d = days; man.= mannitol.

A good response has been obtained with the combination pretreatment of 3 or 4 days in 0.3 M mannitol at 4°C on about 30 genotypes. These include winter genotypes; 'OAC Elmira', GBC776, GBC779, GBC780 and GBC781, and spring genotypes; 'Trinity', 'Cooper', 'Golden Promise', 'Disa', 'Morex', 'Excel', 'Elrose', 'Bruce', 'Rodeo', H936 and 'Sabarilis'. Genotype response failure has not occurred.

Table 3. Effect of time duration that spikes are in 4°C, 0.3 M mannitol pretreatment on embryo formation, cv. Igri

Days in pretreatment	Embryos formed (No.)
0	0
4	9,579
7	9,693
14	8,981
21	9,692
28	10,213

Discussion

We have described a system for isolated microspore culture (IMC) of barley and demonstrated that it is effective across a wide range of barley genotypes. The production of embryo structures leads to a high percentage of regeneration and the percentage of doubled haploids (DH's) produced is also high.

Advantages of IMC vs. anther culture

The most important advantage is the high response across genotypes with the absence of a strong genotype effect. There does not appear to be one factor responsible for this and it would appear to be due to the development of a highly responsive system in which most of the limiting factors have been overcome. The ability to mix the isolated microspores from many spikes also ensures more uniform replication for experiments.

Another advantage is the larger number of plants that can be produced and this is likely related to good embryo formation and the resulting high regeneration frequency of completely fertile doubled haploids. Part of the reason may be the removal of the anther wall, which could result in better nutrient availability to microspores and the removal of the anther physical restrictions.

The isolated microspores can be observed directly and their progress monitored. For this study liquid culture medium is used since it is not possible to observe the microspores under a microscope when on the filter paper. This observation of microspores has been important in observing the induction process from nuclear divisions through embryo development.

The large population of uniformly staged haploid microspores provides better targets for mutation, selection and transformation research. In this aspect, the combined cold plus mannitol pretreatment could be advantageous. Using IMC, transgenic plants from bombarded microspores have been selected using the herbicide resistance gene *Bar* (Jähne *et al.*, 1994; Yao *et al.*, 1997) and we have also selected plants using antibiotic resistance genes and the visible marker 'green flourescent protein' (GFP).

A great deal of time and labour is saved by using spike pretreatment and blender isolation of microspores as compared to removal of anthers for culturing. The use of the filter paper also facilitates the transfer of microspores to different media or treatments. The high frequency of completely fertile doubled haploids eliminates the extra time required to treat plants for chromosome doubling and produces more seed per DH plant. Based on results in the literature, anther culture does not appear to have as high a frequency of DH progeny, but this needs to be studied as few comparisons on the same genotypes have been conducted (Devaux *et al.*, 1996).

The main drawback of IMC has been the difficulty in getting the system to work in many labs. Consistent results require good growth facilities that are pest free and great care in tending the donor plants. The isolation and plating of microspores requires skill, training and practice. It would appear that more care is required in cleanliness of procedures because if one loses a plate due to contamination, a lot of material can be lost.

A number of small critical points have been mentioned in the Procedures and Results sections, such as using refrigerated equipment and solutions during microspore isolation, washing of the filters, wrapping of the plates and density of the microspores on the filter or per ml in liquid media. Also important has been the use of PAA as the auxin in the induction media. One may need to experiment with PAA levels since plant growth can apparently affect the endogenous levels and thus the requirement for auxin. From 2 to 10 mg/l PAA is the suggested range to try. Some labs prefer to use 2,4-D which is a stronger auxin and may be more constant in amount used. However, it should be removed within hours and replaced with PAA in order to get a high frequency of good embryo formation.

Pretreatment is a critical step in the production of large number of embryos and plants from isolated microspore culture. The responses reported in Table 2 are very good but indicate that there may be some genotype differences as discussed. Both donor plant growth condition and microspore stage can influence response and microspore stage should be checked for each genotype since it can vary with the plant morphology that one uses for rough staging. If one plans to use only one method, the combination of cold plus mannitol would be recommended due to its speed, delay of nuclear divisions (unpublished data) and potential to overcome some genotype response differences. It would appear that this combination pretreatment may also produce a higher proportion of green plants from genotypes where albinos are obtained, but this has not been clearly documented. The use of the combined cold plus mannitol pretreatment not only induces large numbers of embryoids but also tends to delay microspore divisions, which could lead to a more synchronized population. This could be valuable in microspore transformation, testing expression of gene constructs, in *in vitro* selection, or other cell uses.

In conclusion, this paper has described an isolated microspore culture system for barley that is largely genotype independent and produces large numbers of embryo structures, which regenerate at high frequencies to provide plants, most of which are completely fertile doubled haploids. Many details are not obvious but are critical to success. The procedures require greater attention to detail and cleanliness than anther culture and practice is essential. Work is progressing well on a similar system for wheat. Unfortunately, most details including pretreatment and media components, require extensive re-evaluation for wheat.

Acknowledgements

The authors are particularly grateful for the financial support from the Natural Sciences and Engineering Research Council of Canada and for the facilities and financial support provided by the Ontario Ministry of Agriculture, Food and Rural Affairs.

References

Davies, D.R. and D.A. Hopwood. (1980). The Plant Genome. The John Innes Charity, Norwich

Devaux, P., M. Zivy, A. Kilian, and A. Kleinhofs, 1996. Doubled haploids in barley. In: Proc. V International Oat Conference and VII International Barley Genetics Symposium. Scoles, G. and B. Rossnagel (Eds.) Univ. of Saskatchewen, Saskatoon. pp. 213-222

Dunwell, J.M., 1996. Microspore culture. In: *In Vitro* Haploid Production in Higher Plants. Vol. 1. Jain, S.M., S.K. Sopory and R.E. Veilleux (Eds.) Kluwer Acad. Pub., Dordrecht. pp. 205-216

Hoekstra, S., M.H. van Zijderveld, F. Heidekamp, and F. van der Mark, 1993. Microspore culture of *Hordeum vulgare* L.: the influence of density and osmolality. Plant Cell Rep. **12**: 661-665

Hu, H., 1997. *In vitro* induced haploids in wheat. In: *In Vitro* Haploid Production in Higher Plants. Vol. 4. Jain, S.M., S.K. Sopory and R.E. Veilleux (Eds.) Kluwer Acad. Pub., Dordrecht. pp.73-97

Hunter, C.P. (1988). Plant regeneration from microspores of barley, *Hordeum vulgare* L. Ph.D. Thesis. Wye College Univ., London

Jain, S.M., S.K. Sopory, and R.E. Veilleux. (Eds.) 1996-1997. *In Vitro* Haploid Production in Higher Plants. Vol. 1-5. Kluwer Acad. Pub., Dordrecht

Jähne, A., D. Becker, R. Brettschneider, and H. Lörz, 1994. Regeneration of transgenic, microspore-derived, fertile barley. Theor.Appl.Genet. **89**: 525-533

Jähne, A. and H. Lörz, 1995. Cereal microspore culture. Plant Sci. **109**: 1-12

Kasha, K.J., Q. Yao, E. Simion, T. Hu, and R. Oro, 1995. Production and applications of doubled haploids in crops. In: Induced Mutations and Molecular Genetics for Crop Improvement. IAEA, Vienna. pp. 23-37

Kasha, K.J., A. Ziauddin, and U-H. Cho, 1990. Haploids in cereal improvement: anther and microspore culture. In: Gene Manipulation in Plant Improvement. II. 19th Stadler Genetics Symposium. Gustafson, J.P. (Ed.) Plenum Press, New York. pp.213-236

Kasha, K. J. (Ed.) 1974. Haploids in Higher Plants, Advances and Potential. Univ. of Guelph, Guelph

Maluszynski, M., I. Szarejko, and B. Sigurbjörnsson, 1996. Haploidy and mutation techniques. In: *In Vitro* Haploid Production in Higher Plants. Jain, S.M., S.K. Sopory and R.E. Veilleux (Eds.) Kluwer Acad. Pub., Dordrecht. pp.67-93

Mordhorst, A.P. and H. Lörz, 1993. Embryogenesis and development of isolated barley (*Hordeum vulgare* L.) microspores are influenced by amount and composition of nitrogen sources in culture media. J.Plant Physiol. **142**: 485-492

Olsen, F.L., 1987. Induction of microspore organogenesis in cultured anthers of *Hordeum vulgare*. The effects of ammonium nitrate, glutamine and asparagine as nitrogen sources. Carlsberg Res.Commun. **52**: 393-404

Roberts-Oehlschlager, S.L. and J.M. Dunwell, 1990. Barley anther culture: pretreatment on mannitol stimulates production of microspore-derived embryos. Plant Cell Tiss.Org.Cult. **20**: 235-240

Scott, P. and R.L. Lyne, 1994. The effect of different carbohydrate sources upon the initiation of embryogenesis from barley microspores. Plant Cell Tiss.Org.Cult. **36**: 129-133

Sopory, S.K. and M. Munshi, 1996. Anther Culture. In: *In Vitro* Haploid Production in Higher Plants. Jain, S.M., S.K. Sopory and R.E. Veilleux (Eds.) Kluwer Acad. Pub., Dordrecht. pp.145-176

Wheatley, W.G., A.A. Marsolais, and K.J. Kasha, 1986. Microspore growth and anther staging in barley anther culture. Plant Cell Rep. **5**: 47-49

Yao, Q.A., E. Simion, M. William, J. Krochko, and K.J. Kasha, 1997. Biolistic transformation of haploid isolated microspores of barley (*Hordeum vulgare* L.). Genome. **40**: 570-581

Ziauddin, A., A. Marsolais, E. Simion, and K.J. Kasha, 1992. Improved plant regeneration from wheat anther and barley microspore culture using phenylacetic acid (PAA). Plant Cell Rep. **11**: 489-498.

Heterosis in crop mutant crosses and production of high yielding lines using doubled haploid systems

M. Maluszynski[1], I. Szarejko[2], P. Barriga[3] and A. Balcerzyk[2]
[1] *Plant Breeding and Genetics Section, Joint FAO/IAEA Division, Vienna, Austria*
[2] *Department of Genetics, Silesian University, Katowice, Poland*
[3] *Austral University, Valdivia, Chile*

Key words: mutant heterosis, hybrid vigour, doubled haploids, hybrid performance, barley, *Hordeum vulgare*, wheat, *Triticum aestivum*, 'F_1-performing' DH lines

Summary

Heterosis appearing in crosses between mutants derived from the same parent variety and crosses of mutants with parent varieties has been observed by many authors for many plant species including such important crops as barley, faba bean, maize, pea, pearl millet, rice, sunflower, sesame, sweet clover, triticale and wheat. Mutant heterosis was reported for crosses of spontaneous mutants, mutants obtained after treatment with various mutagens and recently also for somaclonal variants. The heterotic effects are usually related to an increase in some yield components. There is a lack of correlation between mutation of a particular plant character and appearance of heterotic effect. The yielding performance of a mutant was not correlated with its potential to give yield heterosis in crosses with another mutant, often also a very poor one, or with the parent variety. Poor yielding barley mutants from the collection of semi-dwarf forms of the Department of Genetics, Silesian University gave heterosis in crosses with other mutants or parent varieties for such characters as tillering, grain number and weight per plant. In addition to mutants with deleterious mutations of such characters as chlorophyll synthesis or fasciated stem there were also lines mutated for earliness, semi-dwarfness, low glucoside or high protein and lysine content which gave significant heterosis in crosses. One possible explanation of the phenomenon of mutant heterosis is related to the frequency of mutations induced by chemical and physical mutagens. The appearance and the level of heterotic effect of mutated genes will depend on their interaction with other mutated genes or with genes from the parental genotype. High specific combining ability of mutants giving heterotic effect makes hybrid seed production, based on crosses with defined sources of cytoplasmic or genetic male sterility, unfeasible or even impossible. Doubled haploids provide a unique system to attempt the 'fixing' of hybrid performance in homozygous lines and to avoid the step of hybrid seed production. The assumption on the 'fixability' of hybrid yield in homozygous lines based on reports that large additive genetic variance is responsible for yield heterosis in wheat or barley was proven also for mutant crosses.

M. Maluszynski and K.J. Kasha (eds.),
Mutations, In Vitro and Molecular Techniques for Environmentally Sustainable Crop Improvement, 55–73.

Mutant heterosis

The phenomenon of heterosis appearing in crosses between mutants derived from the same parent variety was already observed by many authors in the early 1940's. The heterotic effect was also observed in crosses of mutants with parent varieties. Mutant heterosis was reported for many plant species including such important crops as barley, faba bean, maize, pea, pearl millet, rice, sunflower, sesame, sweet clover, triticale and wheat (Table 1). It was also described for "model plant species" such as *Antirrhinum majus* (Stubbe and Pirschle, 1999) and *Petunia axillaris* (Gorny, 1977). The explanation of this phenomenon is still far from the final conclusion. The heterotic effect of spontaneous or induced deleterious mutations in the heterozygous stage was described by many leading geneticists as early as the 1940's and reviewed by (Gustafsson, 1946). Monogenic inherited sub-lethal changes such as *albina*, *xantha* or *alboviridis* - chlorophyll mutations in barley, *spectabilis* in *Antirrhinum majus* or anthocyanin factor in pea, when heterozygous, can surpass the original pure line and express significant hybrid vigour of the F_1 plant. The simple Mendelian inheritance of mutated characters lead to the concept of "monohybrid heterosis." However, (Doll, 1966), on the basis of experiments with 22 gamma-rays induced barley mutants, pointed out that the majority of chlorophyll-mutant genes have no, or very little, heterozygous effect on yield heterosis and that overdominant gene action occurs very rarely. Nevertheless, the phenomenon has generated many discussions and is still waiting for a solid explanation in the terms of molecular plant physiology.

The other explanation of this phenomenon is related to the frequency of mutations induced by chemical or physical mutagens. There is still a lack of precise data evaluating the frequency of induced mutations at the DNA level in plant cells. However, the data obtained by independent authors for many loci and a number of crop species indicate a very high frequency of induced mutations. In well described experiments, the mutation frequency of loci responsible for earliness was 3.55×10^{-3} for cotton (Saeed Iqbal *et al.*, 1991) and 3.07×10^{-3} for soybean (Kawai, 1967). Similarly, a high frequency of mutations for semi-dwarfness was observed in many experiments leading to the development of short stature varieties in almost all cereals. Even for loci considered as difficult for mutations, the observed frequency was relatively high and ranged from 2.0×10^{-4} for aphid resistance in cowpea (Pathak, 1991) to 4.0×10^{-4} for signaling mutants in tomato (Lightner *et al.*, 1993). There is no reason to conclude that genes responsible for any yield component will mutate with a frequency other than described above. It seems that the appearance and the level of heterotic effect of mutated genes will depend on their interaction with other mutated genes or with genes from the parental genotype. This concept has support in a relatively high frequency of cross combinations with heterotic effect observed in numerous crosses between mutants of various crops described by many authors (Table 1).

Mutant heterosis was reported for crosses of spontaneous mutants, mutants obtained after treatment with various mutagens (Table 1) and recently also for crosses of parent varieties with somaclonal variants (Bouharmont, personal communication). Among radiation induced mutants there are forms selected in progenies segregated after treatment with X-rays, gamma rays, thermal and fast neutrons. Among chemical mutagens, N-methyl-N-nitrosourea (MNH) and ethyl methanesulfonate (EMS) were mainly used. There is no evidence to support the mutagenic specificity in relation to the discussed phenomenon as all these mutagens are well known as highly efficient in induction of many mutations in plant cells. There is also a lack of correlation between mutations of a particular plant character and appearance of heterotic effect. In addition to mutants with deleterious mutations of such characters as chlorophyll

Table 1. Mutant heterosis in crop plants

Type of cross	Mutants used in crosses (No.)	Type of mutants	Heterotic cross combinations (No.)	Character(s) improved	Yield increase[1] above the better parent (%)	Mutagen	References
Barley							
Mutant x mutant	2	Chlorophyll	1	No. of spikes and grains per plant, kernel weight per plant, seed germination	19	Spontaneous	(Gustafsson, 1946; Gustafsson, 1947; Gustafsson *et al.*, 1950)
Parent x mutant, mutant x mutant	3	Erectoides	6	Number of spikes per plant, length of spike, length of culm	29.9	X-rays	(Nybom, 1950)
Parent x mutant	6	Erectoides	4	Plant weight, grain weight per plant, No. of spikes per plant, 1000 grain weight	23.9	X-rays	(Hagberg, 1953)
Heterozygot mutant	49	Chlorophyll	23	No. of kernels per plant	3.4	Gamma-rays	(Doll, 1966)
Parent x mutant, line x mutant	2	Erectoides, earliness	3	Grain weight per plant	11.0	X-rays	(Gustafsson *et al.*, 1973)
Parent x mutant	1	Earliness, photoperiod insensitive	1	Grain weight and grain number per plant, tillering, culm length, fresh and dry plant weight	13.5[5]	X-rays	(Gustafsson and Dormling, 1972)
Mutant x mutant, parent x mutant	21	Semidwarfness	23	Grain weight per plant	250	MNH	(Maluszynski *et al.*, 1987; Maluszynski *et al.*, 1989; Polok *et al.*, 1997)
Faba bean							
Mutants x mutant	7	Topless	21	Grain yield, podded lateral branches per plant, pods set, biomass	30[3]		(Stelling, 1997)

Type of cross	Mutants used in crosses (No.)	Type of mutants	Heterotic cross combinations (No.)	Character(s) improved	Yield increase[1] above the better parent (%)	Mutagen	References
Maize							
Parent x mutant	5	Degenerative	4	Yield per plot	104	Spontaneous	(Jones, 1945)
Mutant x mutant	22	High protein and lysine content,	12	Yield	46.8	Gamma rays, fast neutrons	(Stoilov and Daskaloff, 1976)
Line x mutant	5	Chlorophyll, colour of endosperm	5	100 kernel weight, grain yield, ear length, plant height	7[3]	EMS, X-rays + EMS	(Dollinger, 1983)
Mutant x mutant	5	Various	6	Leaf area, net assimilation rate, grain yield	35.5[3]	Gamma rays	(Pasztor *et al.*, 1985)
Pea							
Mutant x mutant	16	Fasciated and other	15	Seed number per plant	300	X-rays	(Gottschalk, 1976)
Mutant x mutant	2	Early flowering	1	Seed protein	26.4	X-rays	(Kaul, 1980)
Parent x mutants, mutants x mutants[2]	22	Fasciated and other	22	Seed number per plant, height	200	X-rays	(Lönnig, 1982)
Line x mutant	2	Chlorophyll	5	Seed number per plant, weight of seeds, plant height, *in vitro* callus weight	66.4	???	(Rybtsov *et al.*, 1997)
Pearl millet							
Line x mutant	1	Maintainer	1	Seed yield	34.7[3]	Gamma rays	(Rai *et al.*, 1986)
Rice							
Parent x mutant, line x mutant	1	Semidwarfness	5	Yield per plant, 100 grain weight, No. of productive tillers	97.6	Gamma rays	(Anandakumar and Sree Rangasamy, 1995)
Line x mutant	1	Environment-induced genic male sterility, increased stigma exertion rate	2	Grain weight per plant	8.1[3]	Gamma rays	(Shen *et al.*, 1994)
Line x mutant	2	Large grain, tallness, increased panicle	2	Grain weight per plant, 1000 kernel weight	24.1[3]	Gamma rays *in vitro* mutagenesis	(Shen *et al.*, 1995)

Sunflower							
Mutant x mutant, line x mutant	5	Plastom mutation	16	Seed yield, 1000 seeds weight, calathide diameter	242.7	MNH	(Durneva *et al.*, 1994)
Mutant x mutant, line x mutant	5	Plastom mutation	20	Seed yield, 1000 seeds weight, calathide diameter	368.8	MNH	(Zinovatnaya *et al.*, 1995)
Sesame							
Mutant x mutant	3	Multilocular pods	3	Seed yield	75.8 (60.0[4])	Gamma-rays	(Murty, 1979)
Mutant x mutant mutant x variety	3	Earliness, oil content	13	Seed yield, branches/plant, capsules/ plant, 1000 seeds weight, plant height	31.0	Gamma-rays	(Hoballah, 1999)
Sweet clover							
Mutants x parent	5	Low glucoside	5	Plant weight, main shoot length, number of nodes per shoot, length of side brunches	155.9	X-, gamma-rays, thermal neutrons	(Micke, 1969; Micke, 1976)
Mutants x mutants	6	Low glucoside	5	Plant weight, main shoot length, number of nodes per shoot, length of side brunches	138.9	X-, gamma-rays, thermal neutrons	(Micke, 1969; Micke, 1976)
Triticale							
Mutant x mutant	11	Semidwarfness	23	Grain weight per plant, grain number per plant, culm length	232.5	MNH, dMS	(Grzesik, 1995)
Line x mutant	3	Semidwarfness	21	Grain weight per plot	98	MNH	(Spiss and Goral, 1990)
Wheat							
Parent x mutant	12	Various	8	Plant height; grain number & weight per spike and plant	29.1	Gamma rays	(Xu, 1988)
Line x mutant	5	Semidwarfness, tallness	7	Grain weight per plant, harvest index	71.3	Gamma rays	(Konzak, 1989)
Line x mutant	1	Dwarf	9	Yield per plot	17	Spontaneous	(Gale *et al.*, 1988)

[1]/ maximal yield increase described in grain weight/plant; [2]/ including recombinants from crosses with mutants; [3]/ in relation to commercial hybrid or check variety; [4]/ in relation to parent variety; [5]/ average for various characters. Papers with information on heterosis in relation to the mid-parent were not included in the table.

synthesis or fasciated stem there are also lines mutated for earliness, semi-dwarfness, low glucoside or high protein and lysine content which gave significant heterosis in crosses.

The heterotic effects described in Table 1 were usually related to an increase in some yield components. The yielding performance of a mutant was not correlated with its potential to give yield heterosis in crosses with another mutant, often also a very poor one, or with the parent variety. Poor yielding, semi-dwarf, narrow leafed mutant (437 MG) gave heterosis in crosses with the parent (Mg4170) for such characters as tillering, grain number and weight per plant (Polok *et al.*, 1997). In sunflower, the line described for plastom mutation, in crosses with other mutants or lines, produced heterotic F_1 with a significant increase of seed yield, 1000 seeds weight and calathide diameter (Durneva *et al.*, 1994).

The number of cases where mutant heterosis have been observed in particular crops reflects the intensity of mutation research and the size of mutated germplasm used in crosses. It is not related to specific crop species or even groups of crops. Barley examples dominate in the presented list but it should be noted that it was and still is a "model species" for mutation research in cereals.

To test the conclusions drawn from experiments with various crop species, a large scale, long-term experiment was organized with mutants from the semi-dwarf barley collection of the Department of Genetics, University of Silesia. The ability of mutants to produce "F1-performing" hybrids is demonstrated by the data obtained in the years 1996-1998. The experiment followed the methodology described in previous papers (Maluszynski *et al.*, 1987; Maluszynski *et al.*, 1989). Eighty seven mutants of four barley cultivars, 'Aramir', 'Rudzik', 'Maresi', 'Dema' and doubled haploid mutants from Canadian doubled haploid line H930-36 (derived from the cross of 'Klages' x 'Mata') have been used for crosses in this experiment. All mutants were randomly chosen except for two mutants from the variety Aramir. The Aramir mutants 032AR and 067AR were already described as giving a very high heterotic effect in the F_1 when crossed together (Maluszynski, 1982). These two mutants were crossed with other Aramir mutants to investigate their heterotic potential in various genetic backgrounds. All these mutants were obtained after mutagenic treatment with MNH or after doubled treatment with sodium azide (NaN_3) and MNH (Szarejko and Maluszynski, 1999). Only six characters, mainly related to the agronomic performance of the hybrid (height, spike length, spikes and grain number, grain weight per plant and the 1000 grain weight) were studied.

Numerous mutants from all varieties and DH line H930-36 gave heterosis in the F_1 except for mutants from variety Dema. Among 7 cross combinations with 9 Dema mutants, none gave heterosis in the F_1. Twenty one (24.1%) out of 87 investigated mutants gave heterotic effects in crosses (Table 2). Heterosis in the F_1 was observed in 19 (23.7%) of 80 investigated cross combinations. Heterosis was reported mainly for such characters as grain number (17.5%) or grain weight per plant (17.5%). In 12 (15%) cross combinations heterosis was observed for both these traits. In some cases, heterosis in the described characters was connected with an increased tillering of F_1 (12.5%). However, none of the Maresi mutant cross combinations gave heterosis related to tillering. The heterotic effect in the height of plant was observed only in one case as the semi-dwarf mutants used for this experiment expressed complementation effects for the plant height.

This experiment confirmed previously drawn conclusions that the ability of mutants for building heterotic F_1 in crosses with other mutants or parent varieties is not related to their agronomic performance. Among 21 mutants giving heterosis in crosses only mutant MR207/4 from variety Maresi gave a higher yield in grain number per plant than the parent. Additionally, for the same character one mutant and two DH mutant lines yielded similar to the parent varieties. A decrease in value or lack of mutational changes for such characters as plant height,spike length and spikes per plant was observed in the majority of investigated mutants (Table 3). In characters directly describing plant yield - grain number and weight per plant - 17 mutants indicated a significant decrease.

Previous observations (Maluszynski *et al.*, 1989) on the relatively high frequency of heterotic F_1 from crosses of mutants with their parent varieties was also confirmed (Table 4). The level of heterosis in these crosses was high: 32.6% and 35.5% for crosses with Aramir, 45.1% and 50.7% maximum value for three crosses with Rudzik and 23.0% and 48.0% for a cross with Maresi, for grain number and weight per plant, respectively. The same phenomenon was observed in a cross of the DH mutant AS7 with its parental DH line H930-36. In this case it was clearly demonstrated that through mutagenic treatment of a pure homozygous line (doubled haploid) it was possible to obtain also a completely homozygous line, which is giving heterosis in crosses with the parent. In other words, a near-isogenic line obtained by mutations is able to produce a heterotic hybrid in a cross with the parent. In this case the observed heterosis cannot be attributed to an assumed, but hidden hetrozygosity of the parental material. This statement led to the conclusion that a limited number of genes changed by mutagenic treatment are directly responsible for yield heterosis in the F_1. Results from crosses of two Aramir mutants, which in more then ten years of experiments, indicated hetrosis in many cross combination leads to a similar conclusion. The mutant 067AR expressed a very high combining ability with various mutants from Aramir. In the presented experiment this mutant expresses ability to produce heterotic hybrids in crosses with the parent variety Aramir and also in crosses with six other mutants where it was used as the male or female parent. The maximum value of heterosis in these crosses reached 41.3% for grain and 35.5% for grain weight per plant.

Similar results were obtained for mutant wheat crosses at the Austral University, Chile. Gamma ray induced mutants of advanced generations (M_6-M_8) were used for crosses (Barriga - project report, unpublished data 1997). As in barley, heterosis was observed in crosses of wheat mutants with parental varieties and in crosses among mutants (Table 5). Mutant AU 19 from variety 'Austral' gave a heterotic F_1 in crosses with the parent variety and also with five other Austral mutants. More than 40% of the yield increase in relation to better parents was noted in crosses among mutants. Heterosis was also reported for crosses of three other mutants with their parent varieties.

Gamma ray induced wheat mutants expressed heterosis in crosses with the parent variety 'Zhemai No.3' in such characters as grain number and weight per spike and plant. In some mutant crosses heterosis in plant height was also observed. Twelve mutants were investigated. Heterosis for at least one of the described characters was observed in eight crosses (Xu, 1988).

Heterosis and genetic distance between parents

The presented results from the use of mutants for production of heterotic hybrids concurred with recently published papers related to the prediction of heterosis on the basis of genetic distance estimated by molecular markers polymorphism. It has been the general assumption that the genetic distance between parents is positively correlated with heterosis in the resulting F_1 hybrid but until recently the only data available for testing this hypothesis were based almost entirely on pedigree analysis of the parents. The recent advances in molecular marker technology made it possible to evaluate directly whether the genetic diversity between parents at the DNA level can be used as a tool in predicting heterosis. There are many studies examining this question for such crops as maize, wheat, rice and soybean but the results are

Table 2. Number of combinations with significant heterotic effect in F_1 of barley mutant crosses investigated in years 1995-97

Variety	No. of mutants		No. of cross combinations		No. of combinations with heterotic effect in relation to different characters						
	Used for crosses	Giving heterotic effect	Investigated	With heterotic effect	Plant height	Spike length	Spikes per plant	Grain number per plant	Grain weight per plant	Grain number and weight per plant	1000 grain weight
Aramir	16	9	20	9	-	2	4	6	6	5	2
Rudzik	17	4	18	4	1	-	4	4	4	4	-
Maresi	8	4	10	3	-	2	-	1	2	1	3
Dema	9	-	7	-	-	-	-		-	-	-
DH line H930-36	37	4	25	3	-	-	2	3	2	2	-
Total	87	21 (24.1%)	80	19 (23.7%)	1 (1.25%)	4 (3.7%)	10 (12.5%)	14 (17.5%)	14 (17.5%)	12 (15.0%)	5 (6.3%)

Table 3. Overview of the performance of mutants giving heterotic effect in comparison to the parent variety

Parent genotype	No. of mutants giving heterosis	Number of mutants with or without significant changes in investigated characters																	
		Plant height			Spike length			Spikes per plant			Grain number per plant			Grain weight per plant			1000 grain weight		
		<	=	>	<	=	>	<	=	>	<	=	>	<	=	>	<	=	>
Aramir	9	8	1	-	6	3	-	9	-	-	9	-	-	9	-	-	4	5	-
Rudzik	4	3	1	-	1	3	-	2	2	-	4	-	-	4	-	-	3	1	-
Maresi	4	3	1	-	2	2	-	2	2	-	2	1	1	2	2	-	1	3	-
DH H930-36	4	2	2	-	1	3	-	-	4	-	2	2	-	2	2	-	2	2	-
Total	21	16	5	-	10	11	-	13	8	-	17	3	1	17	4	-	10	11	-

< >/Number of mutants inferior or superior to the parent variety for an investigated character; =/Number of mutants not different from the parent variety for an investigated character; significant differences on the basis of analysis of variance for P=0.05

Table 4. F_1 heterosis in barley mutant crosses

Genotype	Height		Spike length		Spikes per plant		Grain per plant		Grain weight per plant		1000 grain weight	
	(cm)	(%)	(cm)	(%)	(No.)	(%)	(No.)	(%)	(g)	(%)	(g)	(%)
Aramir	74.5	100.0	11.9	100.0	22.0	100.0	366.1	100.0	12.9	100.0	35.1	100.0
010AR	64.2[a]	-13.4	9.8[a]	-17.7	16.8[a]	-23.6	246.1[a]	-32.8	9.3[a]	-27.9	37.7	4.6
012AR	71.1	-2.9	12.1	1.7	15.8[a]	-27.3	253.1[a]	-30.2	8.4[a]	-34.1	33.8	-5.1
013AR	64.6[a]	-13.3	12.0	0.8	9.5[a]	-56.8	126.3[a]	-65.5	2.9[a]	-77.5	22.4[a]	-36.2
032AR	45.4[a]	-39.1	12.6	5.9	6.8[a]	-69.1	96.6[a]	-73.6	2.8[a]	-78.3	27.6[a]	-21.4
035AR	36.9[a]	-50.5	9.4[a]	-21.0	8.6[a]	-60.9	129.3[a]	-64.7	2.9[a]	-77.5	21.9[a]	-37.6
055AR	61.9[a]	-16.9	11.4[a]	-4.2	14.6[a]	-33.6	253.7[a]	-30.7	8.9[a]	-31.0	34.5	-1.7
061AR	60.1[a]	-19.3	10.4[a]	-12.6	16.3[a]	-25.9	243.6[a]	-33.5	8.1[a]	-37.2	33.1	-5.7
067AR	52.8[a]	-29.1	9.0[a]	-24.4	16.8[a]	-23.6	226.6[a]	-38.1	6.1[a]	-52.7	25.4[a]	-26.6
079AR	66.4[a]	-10.9	11.4[a]	-4.2	14.2[a]	-35.4	214.1[a]	-41.5	7.8[a]	-39.5	36.0	-2.6
F_1 067AR x Aramir	71.4	-4.2	11.4[a]	-4.7	25.6	9.8	486.0[b]	32.6	17.5[b]	35.5	35.8	2.0
F_1 067AR x 010AR	72.6	2.9	10.7[a]	-10.1	24.9	13.2	438.7	19.8	16.5[b]	27.9	38.5	9.7
F_1 012AR x 067AR	76.7	2.9	12.1	1.7	31.9[b]	45.0	504.3[b]	37.7	17.7[b]	37.2	34.8	-0.8
F_1 067AR x 035AR	69.9	-6.3	11.9	-0.7	28.8[b]	30.9	464.0[b]	26.8	15.6	20.9	33.6	-4.3
F_1 067AR x 055AR	69.8	-6.3	11.6[a]	-2.8	26.0	18.2	477.1[b]	30.3	17.1[b]	32.5	35.6	1.5
F_1 061AR x 067AR	67.9[a]	-8.8	11.0[a]	-8.0	30.3[b]	37.6	517.3[b]	41.3	17.4[b]	34.5	34.0	-3.1
F_1 079AR x 067AR	70.9	-4.8	11.0[a]	-7.8	27.8[b]	26.3	457.6[b]	24.9	15.9[b]	23.2	34.7	-1.9
F_1 013AR x *032AR*	76.6	2.8	14.5[b]	21.8	17.9	-18.8	362.0	-1.1	14.8	14.9	41.5[b]	18.2
F_1 085AR x *032AR*	77.6	4.1	14.4[b]	20.5	16.7	-24.2	347.2	-5.2	14.4	11.5	41.8[b]	19.2
Rudzik	85.5	100.0	10.9	100.0	19.8	100.0	401.8	100.0	14.2	100.0	36.7	100.0
R133/10	80.4	-5.9	11.0	0.9	16.3	-17.6	259.6[a]	-35.3	7.9[a]	-44.3	30.9	-15.8
R174/7	79.5[a]	-7.0	9.7	-11.0	14.3[a]	-27.7	245.4[a]	-38.9	7.2[a]	-49.2	29.6[a]	-19.3
R174/8	65.5[a]	-23.3	10.0	-8.2	15.1	-23.7	263.9[a]	-34.3	6.2[a]	-56.3	23.3[a]	-36.5
R184/2	53.8[a]	-37.0	7.1[a]	-34.8	4.8[a]	-75.7	22.7[a]	-94.3	0.3[a]	-97.8	13.1[a]	-64.3

Genotype	Height		Spike length		Spikes per plant		Grain per plant		Grain weight per plant		1000 grain weight	
	(cm)	(%)	(cm)	(%)	(No.)	(%)	(No.)	(%)	(g)	(%)	(g)	(%)
F_1 Rudzik x R133/10	84.6	-1.0	11.1	1.8	25.6[b]	29.3	568.0[b]	41.4	21.8[b]	53.7	38.7	4.5
F_1 Rudzik x R174/8	85.2	-0.3	10.9	0.0	32.6[b]	64.6	589.2[b]	46.6	21.0[b]	47.9	35.6	-3.0
F_1 R184/2 x Rudzik	84.3	-1.4	10.8	-0.9	28.7[b]	44.9	556.4[b]	38.5	20.3[b]	42.9	36.5	-0.5
F_1 R184/2 x R174/7	81.9	-4.2	10.9	0.0	26.7[b]	34.8	504.3[b]	25.5	17.4[b]	22.5	34.5	-6.0
Maresi	79.9	100.0	10.8	100.0	13.6	100.0	269.0	100.0	10.2	100.0	37.3	100.0
MR207/4	63.8[a]	-20.1	11.1	2.7	16.1	18.3	330.1[b]	22.7	9.1	-10.7	27.4[a]	-26.5
MR219/4	58.1[a]	-27.2	7.7[a]	-28.7	8.6[a]	-36.7	143.4[a]	-46.6	4.9[a]	-51.9	35.1	-5.8
MR221/4	59.4[a]	-25.6	7.9[a]	-26.8	8.6[a]	-36.7	149.2[a]	-44.5	5.3[a]	-48.0	36.0	-3.4
MR223/3	74.4	-6.8	10.6	-1.8	15.3	12.5	292.4	8.6	11.1	8.8	38.0	1.8
F_1 Maresi x MR207/4	88.3	10.5	13.2[b]	22.2	14.4	5.8	337.7[b]	25.5	15.4[b]	*51.1*	45.6[b]	22.2
F_1 MR219/4 x MR221/4	78.5	-1.7	10.5	-2.7	16.0	17.6	300.0	11.5	15.5[b]	51.9	50.6[b]	35.6
F_1 MR223/3 x MR221/4	89.1[b]	11.5	12.5[b]	15.7	11.8	-13.2	240.9	-10.4	11.6	13.7	47.9[b]	28.4
DH line H930-36	70.2	100.0	7.8	100.0	15.8	100.0	192.4	100.0	6.4	100.0	33.9	100.0
DH mutant AS7	66.4	-5.4	7.7	-1.3	15.1	-4.4	163.6	-15.0	5.1	-20.3	31.3	-7.7
DH mutant AS30	61.4[a]	-12.5	7.1	-9.0	13.6	-13.9	131.7[a]	-31.5	3.1[a]	-51.6	22.8[a]	-32.7
DH mutant BS19	61.0[a]	-13.1	7.0[a]	-10.3	14.8	-6.3	129.6[a]	-32.6	3.5[a]	-45.3	26.8[a]	-20.9
DH mutant M32	66.9	-4.7	8.0	2.5	16.0	1.3	195.4	1.6	6.1	-4.7	32.2	-5.0
F_1 AS7 x H930-36	71.1	1.3	8.2	5.1	18.4	16.5	260.0[b]	35.1	8.5[b]	32.8	32.7	-3.5
F_1 AS7 x AS30	69.3	-1.3	7.6	-2.6	22.1[b]	39.9	243.4[b]	26.5	6.9	7.8	28.3[a]	-13.5
F_1 M32 x BS19	69.8	-0.6	8.3	6.4	23.2[b]	46.8	261.3[b]	35.8	8.7[b]	35.9	33.9	0.0

[a] - significantly lower than parent variety at P = 0.05; [b] - significantly higher than parent variety at P = 0.05

Table 5. F_1 heterosis in gamma ray induced wheat mutant crosses (Valdivia, 1997)

Cross	Height			Yield			100 grain weight		
	(cm)	(% of BP)	(% of PV)	(g/m^2)	(% of BP)	(% of PV)	(g)	(% of BP)	(% of PV)
Austral x Mutant AU 19	94.4	1.5	7.9	808.7	20.6[a]	20.6[b]	4.74	11.0[a]	11.0[b]
Mutants AU 05 x AU 19	104.2	-3.1	19.1[b]	736.1	26.6[a]	9.8	5.26	7.1	23.2[b]
Mutants AU 09 x AU 19	88.8	-4.5	1.5	828.2	42.4[a]	23.5[b]	5.46	20.8[a]	27.9[b]
Mutants AU 14 x AU 19	93.0	0.0	6.3	770.8	29.4[a]	14.9	4.91	4.2	15.0[b]
Mutants AU 22 x AU 19	89.6	-3.7	2.4	913.4	37.5[a]	36.2[b]	5.04	11.0	18.0[b]
Mutants AU 23 x AU 19	96.2	3.4	9.9[b]	816.6	38.3[a]	21.8[b]	4.92	11.4[a]	15.2[b]
Huenufen x Mutant HU 02	103.2	11.0	13.4[b]	635.2	22.5[a]	31.1[b]	4.95	9.0	11.3[b]
Carahue x Mutant CA 01	98.4	-5.4	-5.4	726.6	6.2	19.9[b]	4.42	19.1[a]	24.2[b]
AS x Mutant AS 02	102.1	4.2	4.2	776.2	11.1[a]	25.9[b]	4.99	11.1[a]	13.4[b]

BP - better parent; PV - parent variety[a]; [a] - significant in relation to the better parent; [b] - significant in relation to the parent variety

still far from conclusive. For example, Smith *et al.* (1990) found in maize a direct correlation between heterosis and genetic distance of elite inbred lines based on RFLP studies, while Melchinger *et al.* (1990) and Dudley *et al.* (1991) working with different maize germplasm observed a low correlation between the RFLP distance and F_1 performance. The experiment with 10 elite cultivars of *indica* and *japonica* rice widely used in hybrid rice programmes in China showed a high positive correlation of yield heterosis with genetic distance based on RAPD and SSR markers (Xiao *et al.*, 1996). The correlation was observed within *indica* x *indica* and *japonica* x *japonica* crosses but not for *indica* x *japonica* crosses. On the basis of other studies on molecular marker heterozygosity and hybrid performance in rice it was concluded that the relationship between genetic diversity and heterosis is variable, depending on the rice germplasm used and the complexity of genetic basis of heterosis (Zhang *et al.*, 1996; Saghai Maroof *et al.*, 1997).

In similar studies performed on wheat with different DNA markers (STS, RFLP, RAPD) most yield components of heterotic hybrids were not related to the genetic distance between parents (Martin *et al.*, 1995; Barbosa-Neto *et al.*, 1996; Perenzin *et al.*, 1998). The latter authors stated: "information regarding the genetic diversity of the parental lines do not appear to be helpful for predicting F_1 performance". Similar conclusions were drawn by Manjarrez-Sandoval *et al.* (1997) and Cerna *et al.* (1997) from their studies in soybean and by Chowdari *et al.* (1998) in pearl millet. Furthermore, Bohn *et al.* (1999) found that none of the DNA marker systems studied (RFLP, AFLP and SSR) could be used for predicting progeny variance in crosses of winter wheat varieties. For all seven yield traits analysed they found none associated with any measure of genetic similarity between parents.

These results seem to be in agreement with the phenomenon of heterosis observed in crosses of mutants derived from the same parent variety or mutants and their parents. Compared with varieties of different pedigrees, mutants and their parent variety are similar to near-isogenic lines, differing only in a limited number of loci – those, which were mutated (Szarejko and Gustafson, personal communication). The high level of heterosis present in many mutant crosses appears to be related to the specific interaction of mutated genes rather than to the overall genetic diversity between parents. The data of many authors showing a lack or low level of correlation between hybrid performance and genetic distance seem to support this conclusion.

Fixation of heterosis - doubled haploids

The main limitation in practical use of mutant heterosis is a lack of efficient systems of hybrid seed production for most crop species. Except for maize, where *cms* genes have been identified and widely used, and rice where both 3- and 2-line male sterile systems have been reported (Marburger and Wang, 1988; Chopra, 1996) there is a lack of similar available male sterile germplasm for hybrid seed production in wheat, barley or soybean. The high level of heterosis observed in many mutant crosses is often connected with the interaction of mutated genes in a particular cross combination but not in other crosses of the same mutant. In other words, high specific combining ability of mutants giving heterotic effect makes hybrid seed production, based on crosses with defined sources of cytoplasmic or genetic male sterility, unfeasible or even impossible.

Doubled haploids (DH) provide a unique system for production of completely homozygous lines from a heterozygote in one generation. The efficient protocols of anther/microspore culture or wide hybridization and embryo rescue developed for most cereals have opened the possibility to attempt the 'fixing' of hybrid performance in homozygous lines and to avoid hybrid seed production. The assumption on the 'fixability' of hybrid yield in homozygous lines can be based on early reports concluding that large additive genetic variance is responsible for yield heterosis in wheat (Bush *et al.*, 1971) or barley (Upadhyaya and Rasmusson, 1967; Yap and Harvey, 1971). There were reports on the possibility to select, in segregating progenies (F_5-F_8), uniform lines that yielded as well as the heterotic F_1, but these reports are limited (Aastveit, 1964; Bush *et al.*, 1971; Sarawat *et al.*, 1994; Xiao *et al.*, 1995). The population of doubled haploids obtained from the heterotic F_1 gives a much higher chance to 'fix' heterosis as all DH lines are genetically stable and those having the 'F_1-performing' combination of genes can be identified with a much higher frequency than in segregating generations.

The first attempt to use doubled haploid procedures for fixation of heterosis was undertaken in barley by Kasha and co-workers as early as 1977 (Kasha *et al.*, 1977). Using the 'bulbosum' method, they produced DH lines from 3 F_1 hybrids showing significant yield heterosis over the best parent (cultivars or breeding lines). Even with the limited number of DH lines tested, they were able to identify at least one DH line per cross similar to or higher in yield than the heterotic F_1. Similarly, DH lines significantly surpassing the best parent were obtained from F_1 hybrids expressing high yield heterosis in rice and wheat (Table 6). In the study of Ba Bong and Swaminathan (1995), DH lines were produced through anther culture from the heterotic F_1s of three commercial *indica* rice varieties. The best DH line, although not at the level of F_1 heterosis, outyielded the best parent by 1.1 t/ha in a plot experiment. In the experiment with wheat (Suenaga, 1994), the best DH lines produced by wide hybridization from a cross of distantly related varieties, expressed an increased yield over the best parent even higher than the heterotic hybrid (e.g. 58.3% and 23.8%, respectively). In all these examples cultivars or breeding lines were used as parents of crosses expressing F_1 heterosis. We have applied the same strategy for 'fixation' of heterosis observed in crosses of mutants derived from the same parent variety or mutant with it's parent.

F_1 hybrids of three crosses which have been constantly giving a high level of heterosis in long term experiments (Maluszynski *et al.*, 1987; Maluszynski *et al.*, 1989) were used as donors for anther culture. In all three cross combinations, 'F_1-performing' DH lines (i.e. surpassing the yield of the parent variety and not significantly different from the heterotic hybrid), were obtained with the relatively high frequency of 1.5-14.3% per cross (Polok *et al.*, 1997). The experiment was continued and stability of DH lines confirmed in subsequent years. For example, DH line D57 derived from the cross of Diva and its mutant 228DV as well as two DH lines M3 and M43 from the cross of mutant 437 MG and its parent Mg 4176 have approached the yield of the heterotic F_1 and surpassed their parent variety by 36.6%, 46.0% and 70.0 %, respectively (grain weight per plant). Additionally, some of the high yielding DH lines were used as parents for the next round of crosses and their F_1 was evaluated together with the initial hybrids, mutants and their parent variety. Examples of results from 1998 are presented in Table 7. High yielding DH lines confirmed their performance and some new lines, better than the parent variety, have been selected. The F_1 hybrid produced from the cross of two high yielding DH lines (M3 x M43) has reached, although not surpassed, the yield of the initial heterotic F_1 hybrid.

Table 6. High yielding doubled haploid lines developed from heterotic hybrids

Species	Cross and generation	Height (cm)	Spike/panicle number	Grain number	Grain weight (g)	Yield t/ha	References
Barley	Diva x Mutant 228DV		per plant	per plant	per plant		(Polok *et al.*, 1997)
	P_1	85.3*	16.3*	362.0	14.5	4.80	
	P_2	76.5*	15.1*	291.5	10.9	4.35	
	F_1	87.5*	25.1*	534.6	23.1	-	(* unpublished data)
	The best DH line	80.3*	23.7*	483.4	19.2	5.65	
Rice	Pusa 743-1-1 x IR66		per plant	per panicle	1000 grain weight		(Ba Bong and Swaminathan, 1995)
	P1	114	6.7	96	29.8	5.5	
	P2	102	8.8	94	21.9	5.2	
	F1	105	9.9	125	23.5	7.0	
	The best DH line	97	12.8	102	31.6	6.6	
Rice	Pusa 743-1-1 x IR72						(Ba Bong and Swaminathan, 1995)
	P_1	114	6.7	96	29.8	5.5	
	P_2	105	9.3	91	22.4	5.3	
	F_1	108	12.4	138	26.5	7.8	
	The best DH line	124	12.8	98	27.2	6.6	
Rice	IR70 x Khonorullo				per plant		(Gupta *et al.*, 1996)
	P_1	57	5.0	123	0.4	3.1	
	P_2	129	4.0	145	5.2	2.2	
	F_1	85	7.0	147	8.1		
	The best DH line	87	3.8	203	7.9	4.0	
Wheat	Fukuho-komugi x Oligo Culm	Culm length	per m^2		per spike		(Suenaga, 1994)
	P_1	93.1	332		1.32	4.37	
	P_2	93.3	259		1.31	3.38	
	F_1	109.7	327		1.65	5.41	
	The best DH line	96.0	397		1.74	6.92	

Table 7. F_1-performing doubled haploids from heterotic barley mutant crosses (1998)

Genotype	Height		Spikes per plant		Grains per plant		Grain weight per plant	
	(cm)	(%)	(No.)	(%)	(No.)	(%)	(g)	(%)
Diva	79.9	100.0	22.4	100.0	396.8	100.0	16.8	100.0
Mutant 228DV	66.3[a]	-16.9	18.8	-15.9	282.6	28.8	10.5	-37.4
F_1 Diva x 228DV	79.0	-1.1	31.1[b]	39.1	531.2[b]	33.9	22.6[b]	35.1
DH line D57	79.2	-0.8	30.8[b]	37.7	547.5[b,c]	38.0	22.9[b,c]	36.6
Mg4170	71.8	100.0	18.4	100.0	192.9	100.0	7.6	100.0
Mutant 437MG	56.9[a]	-20.7	14.1	-23.3	133.3	-30.9	3.7[a]	-51.2
F_1 437MG x Mg4170	82.4[b]	14.8	28.6[b]	55.2	414.1[b]	115.0	16.4[b]	116.7
DH line M3	75.0	4.5	25.0	35.4	338.0[b,c]	75.2	13.6[b,c]	79.1
DH line M43	74.7	4.1	23.7	28.4	320.8[b,c]	66.3	11.1[b]	46.0
F_1 DH-M43 x DH-M3*	79.6[b]	10.9	30.1[b,c]	63.6	359.6[b,c]	86.4	12.9[b]	70.0

[a] - significantly lower than parent variety at P=0.05; [b] - significantly higher than parent variety at P=0.05; [c] - not significantly different from F_1 mutant x parent variety at P=0.05; * - data from two replications

Conclusions

- The phenomenon of heterosis in crosses between mutants derived from the same parent variety or mutants with their parent varieties is common for many plant species including important cereals, legumes and oil crops.
- The heterotic effect is often manifested by an increase of some yield components in mutant hybrids as compared to the parent variety.
- There is a lack of correlation between agronomic performance of mutants and appearance of heterotic effect in their crosses. Both, well performing mutants or mutants with deleterious characters, can give high yield heterosis in F_1 crosses with another mutant or the parent variety.
- Mutants and their parent variety are similar to near-isogenic lines, differing only in a limited number of loci – those, which were mutated. The high level of heterosis expressed in many mutant crosses appears to be related to the specific interaction of mutated genes rather than to the overall genetic diversity between parents.
- Doubled haploids provide a unique system to attempt the 'fixation' of hybrid performance in homozygous lines and to avoid all difficulties related to the production of hybrid seeds. It was possible to produce 'F_1-performing' high yielding doubled haploid lines from the heterotic F_1s of barley, wheat and rice.
- The 'fixability' of mutant heterosis in DH lines is especially promising, as high specific combining ability of mutants giving heterotic effect makes hybrid seed production, based on crosses with defined sources of male sterility, difficult or unfeasible. The use of the DH system allows for wide exploitation of the mutant heterosis phenomenon in breeding of new, high yielding varieties of crops with a well established doubled haploid production system.

References

Aastveit, K., 1964. Heterosis and selection in barley. Genetics. **49**: 159-164

Anandakumar, C.R. and S.R. Sree Rangasamy, 1995. Heterosis and selection indices in rice. Egypt.J.Genet.Cytol. **14**: 123-132

Ba Bong, B. and M.S. Swaminathan, 1995. Magnitude of hybrid vigour retained in double haploid lines of some heterotic rice hybrids. Theor.Appl.Genet. **90**: 253-257

Barbosa-Neto, J.F., M.E. Sorrells, and G. Cisar, 1996. Prediction of heterosis in wheat using coefficient of parentage and RFLP-based estimates of genetic relationship. Genome. **39**: 1142-1149

Bohn, M., H.F. Utz, and A.E. Melchinger, 1999. Genetic similarities among winter wheat cultivars determined on the basis of RFLPs, AFLPs, and SSRs and their use for predicting progeny variance. Crop Sci. **39**: 228-237

Bush, R.H., A. Lucken, and R.C. Frohberg, 1971. F_1 hybrids versus random F_5 line performance and estimates of genetic effects in spring wheat. Crop Sci. **11**: 357-361

Cerna, F.J., S.R. Cianzio, A. Rafalski, S. Tingey, and D. Dyer, 1997. Relationship between seed yield heterosis and molecular marker heterozygosity in soybean. Theor.Appl.Genet. **95**: 460-467

Chopra, K., 1996. Global assessment of hybrid rice technologies for breeding and seed production. In: Proceedings of the 18th Session of the International Rice Commission. Tran, D.V. (Ed.) FAO, Rome. pp.23-36

Chowdari, K.V., S.R. Venkatachalam, A.P. Davierwala, V.S. Gupta, P.K. Ranjekar, and O.P. Govila, 1998. Hybrid performance and genetic distance as revealed by the (GATA)4 microsatellite and RAPD markers in pearl millet. Theor.Appl.Genet. **97**: 163-169

Doll, H., 1966. Yield and variability of chlorophyll-mutant heterozygotes in barley. Hereditas. **56**: 256-276

Dollinger, E.J., 1983. Effects of visible recessive alleles on vigor characteristics in a maize hybrid. Crop Sci. **25**: 819-821

Dudley, J.W., M.A. Saghai Maroof, and G.K. Rufener, 1991. Molecular markers and grouping of parents in maize breeding programs. Crop Sci. **31**: 718-723

Durneva, G.N., I.I. Ulitcheva, Y.D. Beletskii, and E.P. Gus'kov, 1994. Influence of a mutant plastom on the expression of morphological and physiological traits of sunflower hybrids. Russian J.Genet. **30**(4): 437-443

Gale, M.D., A.M. Salter, F.C. Curtis, and W.J. Angus, 1988. The exploitation of the Tom Thumb dwarfing gene, *Rht3*, in F_1 hybrid wheats. In: Semi-dwarf Cereal Mutants and Their Use in Cross-Breeding III. IAEA, Vienna. pp.57-68

Gorny, A. (1977) Chemomutanty *Petunia axillaris* (Lam.) B.S.P. Silesian University, Katowice.

Gottschalk, W., 1976. Monogenic heterosis. In: Induced Mutations in Cross-Breeding. IAEA, Vienna. pp.189-197

Grzesik, H., 1995. Studies on dwarf mutants of winter triticale (X *Triticosecale* Witt.). Part II. Effect of heterosis in F_1 hybrids of winter triticale Hod.Rosl.Aklim.Nasien. **39**(3): 21-39 (in Polish with English summary)

Gupta, H.S., R.N. Bhuyan, A. Pattanayak, and D.K. Pandey, 1996. Development of cold-tolerant rice through anther culture. IRRN. **21**(1): 20

Gustafsson, A., 1946. The effect of heterozygosity on viability and vigour. Hereditas. **32**: 263-286

Gustafsson, A., 1947. The advantageous effect of deleterious mutations. Hereditas. **33**: 573-575

Gustafsson, A. and I. Dormling, 1972. Dominance and overdominance in phytptron analysis of monohybrid barley. Hereditas. **70**: 185-216

Gustafsson, A., I. Dormling, and G. Ekman, 1973. Phytotron ecology of mutant genes. I. Heterosis in mutant crossings of barley. Hereditas. **74**: 119-126

Gustafsson, A., N. Nybom, and U. von Wettstein, 1950. Chlorophyll factors and heterosis in barley. Hereditas. **36**: 383-392

Hagberg, A., 1953. Heterozygosity in *erectoides* mutations in barley. Hereditas. **39**: 161-178

Hoballah, A.A., 1999. Selection and agronomic evaluation of induced mutant lines of sesame. In: Induced Mutations for Sesame Improvement. **IAEA-TECDOC-.** IAEA, Vienna. pp.71-84

Jones, D.F., 1945. Heterosis resulting from degenerative changes. Genetics. **30**: 527-542

Kasha, K.J., L.S.P. Song, S.J. Park, and E. Reinbergs, 1977. Fixation of heterosis: comparison of F_1 hybrids with their respective homozygous lines developed using doubled haploid procedures. Cereal Res.Commun. **5**(3): 205-214

Kaul, M.L.H., 1980. Radiation genetics studies in garden pea. 9. Non-allelism of early flowering mutants and heterosis. Z.Pflanzenzüchtg. **84**: 192-200

Kawai, T., 1967. New crop varieties bred by mutation breeding. Jpn.Agric.Res.Q. **2**: 8-12

Konzak, C.F., 1989. Exploitation and analysis of heterosis in wheat with induced mutations. In: Current Options for Cereal Improvement. Maluszynski, M. (Ed.) Kluwer Academic Publishers, Dordrecht. pp.97-113

Lightner, J., G. Pearce, C.A. Ryan, and J. Browse, 1993. Isolation of signaling mutants of tomato (*Lycopersicum esculentum*). Mol.Gen.Genet. **241**: 595-601

Lönnig, W.-E., 1982. Dominance, overdominance and epistasis in *Pisum sativum* L. Theor.Appl.Genet. **63**: 255-264

Maluszynski, M., 1982. The high mutagenic effectiveness of MNUA in inducing a diversity of dwarf and semi-dwarf forms of spring barley. Acta Soc.Bot.Polon. **51**: 429-440

Maluszynski, M., A. Fuglewicz, I. Szarejko, and A. Micke, 1989. Barley mutant heterosis. In: Current Options for Cereal Improvement. Doubled Haploids, Mutants and Heterosis. Kluwer Academic Publisher, Dordrecht. pp.129-146

Maluszynski, M., A. Micke, B. Sigurbjörnsson, I. Szarejko, and A. Fuglewicz, 1987. The use of mutants for breeding and for hybrid barley. In: Barley Genetics V. Yasuda, S. and T. Konishi (Eds.)Okayama. pp.969-977

Manjarrez-Sandoval, P., T.E. Jr. Carter, D.M. Webb, and J.W. Burton, 1997. Heterosis in soybean and its prediction by genetic similarity measures. Crop Sci. **37**: 1443-1452

Marburger, J.E. and R.R.C. Wang, 1988. Anther culture of some perennial triticeae. Plant Cell Rep. **7**: 313-317

Martin, J.M., L.E. Talbert, S.P. Lanning, and N.K. Blake, 1995. Hybrid performance in wheat as related to parental diversity. Crop Sci. **35**: 104-108

Melchinger, A.E., M. Lee, K.R. Lamkey, and W.L. Woodman, 1990. Genetic diversity for restriction fragment length polymorphisms among maize inbreds with agronomic performance of their crosses. Crop Sci. **33**: 944-950

Micke, A., 1969. Improvement of low yielding sweet clover mutants by heterosis breeding. In: Induced Mutations in Plants. IAEA, Vienna. pp.541-550

Micke, A., 1976. Hybrid vigour in mutant crosses. Prospects and problems of exploitation studied with mutants of sweet clover. In: Induced Mutations in Cross-Breeding. IAEA, Vienna. pp.199-218

Murty, G.S.S., 1979. Heterosis in inter-mutant hybrids of *Sesamum indicum* L. Curr.Sci. **48**(18): 825-827

Nybom, N., 1950. Studies on mutations in barley. I. Superdominant factors for internode length. Hereditas. **36**: 321-328

Pasztor, K., P. Pepo, and K. Egri, 1985. Changes in the production of maize hybrids due to mutant parent lines. Acta.Agron.Acad.Hungar. **34**(1-2): 189-195

Pathak, R.S., 1991. Genetic evaluation of two aphid resistant cowpea mutants in Kenya. In: Plant Mutation Breeding for Crop Improvement. Vol. 2. IAEA, Vienna. pp.241-247

Perenzin, M., M. Corbellini, M. Accerbi, P. Vaccino, and B. Borghi, 1998. Bread wheat: F_1 hybrid performance and parental diversity estimates using molecular markers. Euphytica. **100**: 273-279

Polok, K., I. Szarejko, and M. Maluszynski, 1997. Barley mutant heterosis and fixation of 'F_1 - performance' in doubled haploid lines. Plant Breed. **116**: 133-140

Rai, K.N., D.J. Andrews, and A.S. Rao, 1986. Hybrid potential of 81A - a dwarf male-sterile line of pearl millet. Indian J.Genet. **46**(2): 295-303

Rybtsov, S.A., T.A. Ezhova, and S.A. Gostimskii, 1997. Use of *in vitro* tissue culture to study heterosis in the pea. Russian J.Genet. **33**(11): 1299-1303

Saeed Iqbal, R.M., M.B. Chaudhry, M. Aslam, and A.A. Bandesha, 1991. Economic and Agricultural impact of mutation breeding in cotton in Pakistan. In: Plant Mutation Breeding for Crop Improvement. Vol. 1. IAEA, Vienna. pp.187-201

Saghai Maroof, M.A., G.P. Yang, Z. Qifa, and K.A. Gravois, 1997. Correlation between molecular marker distance and hybrid performance in U.S. southern long grain rice. Crop Sci. **37**: 145-150

Sarawat, P., F.L. Stoddard, and D.R. Marshall, 1994. Derivation of superior F5 lines from heterotic hybrids in pea. Euphytica. **73**: 265-272

Shen, Y., Q. Cai, M. Gao, and Z. Liang, 1995. Isolation and genetic characterization of somaclonal mutants with large-sized grain in rice. Cereal Res.Commun. **23**(3): 235-241

Shen, Y., M. Gao, and Q. Cai, 1994. A novel environment-induced genic male sterile (EGMS) mutant in *indica* rice. Euphytica. **76**: 89-96

Smith, O.S., J.S.C. Smith, S.L. Bowen, R.A. Tenborg, and S.J. Wall, 1990. Similarities among a group of elite maize inbreds as measured by pedegree, F_1 grain yield heterosis and RFLPs. Theor.Appl.Genet. **80**: 833-840

Spiss, L. and H. Goral, 1990. Effect of heterosis in hybrids with dwarf mutants of triticale. Hod.Rosl.Aklim.Nasien. **34**(3/4): 33-37

Stelling, D., 1997. Heterosis and hybrid performance in topless faba beans (*Vicia faba* L.). Euphytica. **97**: 73-79

Stoilov, M. and S. Daskaloff, 1976. Some results on the combined use of induced mutations and heterosis breeding. In: Induced Mutations in Cross-Breeding. IAEA, Vienna. pp.179-188

Stubbe, H. and K. Pirschle, 1940. Über einen monogen bedingten Fall von Heterosis bei *Antirrhinum majus*. Ber.Deutsche.Bot.Ges. **58**: 546-558

Suenaga, K., 1994. Doubled haploid system using the intergeneric crosses between wheat (*Triticum aestivum*) and maize (*Zea mays*). Bull.Natl.Inst.Agrobiol.Resour. **9**: 83-139

Szarejko, I. and M. Maluszynski, 1999. High frequency of mutations after mutagenic treatment of barley seeds with Na N_3 and MNH with application of interincubation germination period. MBNL. **44**

Upadhyaya, B.R. and D.C. Rasmusson, 1967. Heterosis and combining ability in barley. Crop Sci. **7**: 644-647

Xiao, J., J. Li, L. Yuan, S.R. McCouch, and S.D. Tanksley, 1996. Genetic diversity and its relationship to hybrid performance and heterosis in rice as revealed by PCR-based markers. Theor.Appl.Genet. **92**: 637-643

Xiao, J., J. Li, L. Yuan, and S.D. Tanksley, 1995. Dominance is the major genetic basis of heterosis in rice as revealed by QTL analysis using molecular markers. Genetics. **140**: 745-754

Xu, M., 1988. The hereditary tendency of characters in F_1 generation of wheat crosses with mutants. Acta Agri.Nucl.Sinica. **2**(1): 5-12

Yap, T.C. and B.L. Harvey, 1971. Heterosis and combining ability of barley hybrids in densely and widely seeded conditions. Can.J.Plant Sci. **51**: 115-122

Zhang, Q., Z.Q. Zhou, G.P. Yang, C.G. Xu, K.D. Liu, and M.A. Saghai Maroof, 1996. Molecular marker heterozygosity and hybrid performance in indica and japonica rice. Theor.Appl.Genet. **93**: 1218-1224

Zinovatnaya, G.N., I.I. Ulitcheva, and E.P. Gus'kov, 1995. Combining ability of sunflower plastom mutant lines and expression of heterosis for quantitative traits in intermutant hybrids. Russian J.Genet. **31**(12): 1423-1429.

Haploidy breeding and mutagenesis for drought tolerance in wheat

A.J. Khan, S. Hassan, M. Tariq and T. Khan
Nuclear Institute for Food and Agriculture (NIFA), Tarnab-Peshawar, P.O. Box 446, Pakistan

Key words: anther culture, DH-mutants, doubled haploids, gamma rays, mutagenesis, mutants, *Triticum aestivum*

Summary

Several intraspecific crosses between known drought tolerant wheat varieties and stable high yielding recombinants were made with the objective to develop improved cultivars for the moisture stressed rainfed areas of Pakistan. Five of these crosses were selected for further creation of useful mutations through the application of low doses of gamma rays and development of doubled haploids through anther culture. Anther culture response of the selected irradiated F_1 generations was studied on liquid and solid induction media. The highest number of calli among almost all crosses was produced on Potato-2 induction medium. All the crosses varied greatly in response to callus induction and maximum calli (75%) were obtained from Lyl-73/vee's' cross. Similarly, genotypic differences were found for green vs. albino regenerants. The highest number of green plantlets (12.1%) was recorded for Lu-26/3062. From the developed doubled haploid population 25 DH-mutants were initially selected and nine lines were finally included in multi-locational field tests. Two DH-mutants (i.e. DHML-50 and DHML-9) have potential for better grain yield, earliness, disease resistance and moisture stress tolerance.

M. Maluszynski and K.J. Kasha (eds.),
Mutations, In Vitro and Molecular Techniques for Environmentally Sustainable Crop Improvement, 75–82.

Introduction

It has been estimated that "about one-third of the world's potentially viable land suffers from an inadequate supply of water, and on most of the remainder, crop yields are periodically reduced by drought" (Kramer, 1980). Most of the world's major wheat growing areas is threatened routinely by drought. In Pakistan, wheat yield, in addition to other factors, is badly handicapped by water stress, because the dominant feature of Pakistan's climate is its aridity. A combination of low rainfall, low atmospheric humidity and high solar radiation results in large parts of the country experiencing a vegetative moisture deficit over several months of the year. The water limiting factor is more pronounced in the rainfed areas of the country, which constitute nearly $1/3^{rd}$ of the total wheat acreage. In the North West Frontier Province 62% of the area cultivated to wheat is totally dependent on rainfall, and the pattern and distribution of which is also erratic. Low productivity is the common feature of these areas, which is the result of inadequate and unpredictable rainfall pattern, non-availability of the potential cultivars and socio-economic constraints.

Conventional plant breeding is being employed to develop varieties resistant to stresses, in particular to moisture stress, but progress has been slow. There is a great need to exploit all genetic variability that can be used in breeding for stressful environments. Use of unconventional techniques such as doubled haploid (DH) breeding through anther culture will become more useful in the speeding up of the application of conventional plant breeding methods. The method offers unusual potential for reducing the time required for the production of homozygous progeny from crosses as well as to improve the efficiency by which induced mutants can be recovered from mutagenized seeds or plants (Henry *et al.*, 1982; Ouyang, 1986; Hu, 1986). In China, pure breeding diploid lines have been developed from F_1 or F_2 generations, following anther culture *in vitro* to produce haploid materials for chromosome doubling and so far, 12 new wheat varieties have been developed (Hu, 1996). Most of these lines exhibit wide adaptability, high resistance to drought and infertile soil, and luxuriant growth.

Considering the importance of doubled haploid breeding, irradiated F_1 generations were investigated for their androgenic ability and in succession to produce doubled haploid mutant lines for high yield, drought tolerance and early maturity for the moisture stress rainfed environments.

Material and methods

Genotypes used for hybridization

Different wheat cultivars maintained at the institute under the rainfed wheat improvement programme were crossed with each other in an appropriate fashion in order to develop high yielding, drought tolerant and early maturing wheat lines. The successful crosses were harvested and threshed. Five crosses i.e. Lu-26/3062, C-306/3013, Lu-26/3063, Lyl-73/Vee's' and Lu-26/3066 were selected for further use to induce mutations and haploidy breeding.

Irradiation of F_1 generations

F_1 seed of the selected five recombinant lines was irradiated with a low dose of 150 Gy gamma rays with the objective to create genetic variability and accelerate the androgenic response of these plant material. Approximately 250-300 F1 seeds were irradiated for each cross combination. Seed of each irradiated F_1 was planted separately under normal growth conditions.

Culture media

Two different culture media (Potato-2 liquid medium and MS-modified solid medium) were used for induction of callus from the anthers of irradiated wheat hybrids. Potato-2 induction medium was prepared in the way explained by Chuang *et al.* (1978). However, MS-modified medium was prepared in the way suggested by K.J. Kasha (Canada) during the 1st FAO/IAEA Research Coordination Meeting of this CRP. In this case the MS-medium was supplemented with 90 g/l maltose, 750 mg/l glutamine, 2 mg/l BAP and 2 mg/l 2, 4-D.

Anther culture of irradiated F_1 generations

For anther culture, spikes from each irradiated hybrid genotype were selected on the basis of morphological examination at the time when the flag leaf was fully developed. The inflorescence in a compact leaf sheath was sterilized and then removed aseptically from the sheath. Proper anther stage was checked cytologically before inoculation. Anthers at uninucleate stages (i.e. mid to late uninucleate) from each spike were placed on small petri dishes containing two types of induction media. Approximately 30 anthers were placed on the medium in each petri dish. The dishes were sealed with parafilm, wrapped in aluminium foil and were placed in the incubator at 28°C. Upon the formation of calli data was recorded.

Development of DH-mutant lines from irradiated F_1 generations (M_1F_1)

Calli of 2-3 mm diameter, derived from anthers of different recombinants, were shifted to the agar solidified 190-2 regeneration medium. The regenerants obtained were examined for their ploidy status. Number of green and albino plants was recorded. The albino plants were discarded, however, the green regenerants were treated with a solution of colchicine (0.2%) and dimethyl sulfoxide (DMSO 1.5%) for chromosome doubling.

Agronomic evaluation of the potential DH-mutants under moisture stress

The newly developed potential DH-mutant lines were subjected for agronomic evaluation in a non-replicated nursery. Moisture stress was applied at the seedling, pre-anthesis, and grains filling stages for one-week interval each. Data was recorded for some important agronomic traits and disease susceptibility and finally desirable selections were made.

Multi-locational field testing of selected DH-mutants

Nine promising DH-mutant lines along with a commercial check were field evaluated at three moisture stress locations in 6 rows of 5 meters long plots for each line with a RCB- factorial design containing 4 replications. In addition to visual observations for morphological traits, grain yield and maturity data were recorded for all the tested entries. Statistical analysis was performed using PROC GLM (SAS Institute, 1990).

Proline determination

Proline was extracted from all the selected DH-mutants, including the check variety at the time of exposure to moisture stress. The extraction and measurement of proline was carried out using the method of Singh *et al.* (1973ab).

Results and Discussion

Culture ability of irradiated F_1 generations

Obvious differences of all the irradiated F_1 generations and culture media were observed for responding anthers as well as yields of induced calli (Table 1). Irradiated F_1 at serial No.5 showed better callogenetic potential on either of the culture medium, however, 75% of the anthers produced calli on liquid medium and 26% only on solid medium. This confirms the high effectiveness of wheat anther culture on liquid induction medium. The results confirm those of Konzak and Zhou (1991). However, in contrast the irradiated F_1 S.No.4 expressed better efficiency for callus induction on solid medium. Ding *et al.*, (1991) observed accelerating androgenic response through the use of low doses of gamma rays for F_1 generations. Zapata and Aldemita (1986) demonstrated that minimal stress on the explant by irradiation may not induce irreversible genotypic changes but can stimulate callus induction and plant regeneration.

Table 1. Anther culture response of irradiated wheat hybrids on liquid and solid induction media

Serial No.	Irradiated F_1	Anthers plated (No.)	Responding anthers (No.)	Produced calli (No.)	Anther response (%)	Calli per 100 anthers
			Liquid medium (Potato-2)			
1	Lu26/3062	875	295	345	33.7	39.4
2	Lu26/3063	610	189	290	30.9	47.5
3	Lu26/3066	1045	625	701	59.8	67.1
4	C-306/3013	598	45	24	7.5	4.0
5	Lyl-73/Vee"S"	1215	790	920	65.0	75.7
			Solid medium (MS modified)			
1	Lu26/3062	527	115	63	21.8	11.9
2	Lu26/3063	490	80	46	16.3	9.4
3	Lu26/3066	684	152	90	22.2	13.1
4	C-306/3013	603	118	125	19.6	20.7
5	Lyl-73/Vee"S"	810	305	218	37.6	26.9

Regeneration ability of the irradiated F_1 generations

Although regenerants were obtained from almost all irradiated crosses, the ratio of albino was much higher than green plantlets (Table 2). The high rate of albino's might be due to high temperature at the time of callus formation. Ouyang *et al.*, (1983) stated that up to 100% of regenerated plantlets may be albinos. It has been observed that highest number of albinos was produced by cross Lyl-73/vee's', whereas the maximum number of green plantlets per 100 calli was obtained from Lu-26/3062. At maturity seed was recovered from only 76 regenerants.

Table 2. Regeneration response of irradiated F_1 generations

S. No.	Irradiated F_1	Calli plated (No.)	Regenerants obtained			Green plants per 100 calli
			Green	Albino	Total	
1	Lu26/3062	205	25	36	61	29.7
2	Lu26/3063	175	13	22	35	20.0
3	Lu26/3066	350	21	59	80	22.8
4	C-306/3013	125	15	5	20	16.0
5	Lyl-73/Vee"S"	625	45	68	113	18.1

Agronomic evaluation of the potential DH-mutants under moisture stress

Data on 25 potential DH-mutants in a non-replicated nursery revealed that all the lines produced heads 2-12 days earlier than the commercial check (Inqilab). Variations for plant, tillers per plant, spike length, number of grains per spike and grain yield was observed among all the tested lines (Table 3). Nine DH-mutant lines were selected not only on the basis of higher yield but also having better potential for other parameters. These selected lines have also shown resistance to yellow and leaf rusts.

Multi-locational field testing of selected DH-mutants

Results of multi-locational yield trials revealed that DH-mutants, locations and the two-way interaction (mutants x locations) were statistically significant at 5% level of probability (Table 4). Regarding grain yield potential, DHML-50 and DHML-9 were consistently higher yielding than other recombinants at all the three locations. Although DHML-50 had higher yield at Tarnab, it was at par with DHML-9 at the other two locations. The lowest yield was recorded for DHML-20 at Karak and Tarnab. With regards to maturity DHML-27 was found to be early maturing and might, for this reason, be low yielding at Karak and K.I. Khan. Grains with higher 1000-grain weight were observed in case of check variety followed by DHML-7 and DHML-27 respectively.

Proline accumulation in DH-mutants

Prominent differences were observed among all the recombinant lines in their capacity to accumulate proline under moisture stress (Table 5). The highest concentration of proline was recorded from DHML-9, followed by DHML-12, DHML-27 and DHML-50 respectively. The accumulation of free proline in water stressed leaves has been observed in many plant species (Stewart and Hanson, 1980) including wheat (Tan and Halloran, 1980). The proline profile of some DH-mutants when compared with grain yield showed a positive correlation while in

Table 3. Agronomic performance of selected DH-mutants under moisture stress

DH mutant--source	Days to heading	Plant height (cm)	Tillers/ plant (No.)	Spike length (cm)	Spikelets/ spike (No.)	Grains/ spike (No.)	Grain yield/ plot (g)	Disease (ACI) *	
								Yr	Lr
DHML1--Lu26/3063	121	79.8	5	8.7	16	42	91.8	0.2	0
DHML2--C306/3013	120	100.0	4	7.7	13	33	91.2	0	0
DHML3--C306/3013	120	102.2	6	8.7	17	39	92.4	0	0
DHML6--Lu26/3066	127	82.0	5	9.7	17	39	92.4	0	0
DHML7--Lyl73/Vee"S"	128	112.6	4	9.7	17	45	122.4	0	0
DHML9--Lu26/3066	121	87.6	4	8.7	17	52	121.8	0	10
DHML12--Lu26/3062	119	103.0	5	8.0	15	31	121.2	0.7	0
DHML13--Lu26/3066	126	90.2	4	11.3	15	49	151.8	0.2	0
DHML14--Lu26/3063	122	85.4	6	8.7	16	46	63.00	0	0
DHML15--C306/3013	120	99.0	4	9.7	18	46	91.8	2	0
DHML20--Lyl73/Vee"S"	122	120.2	4	10.3	19	30	123.0	0.2	10
DHML22--C306/3013	121	91.8	5	8.0	17	39	90.0	0	0.2
DHML27--Lu26/3062	120	94.4	4	6.3	14	33	121.8	0	0.2
DHML30--Lu26/3066	119	95.0	3	9.0	15	39	88.2	0	0
DHML33--Lyl73/Vee"S"	128	103.6	3	8.3	17	42	90.6	0	0
DHML35--Lu26/3063	129	79.2	4	10.0	18	38	61.2	0.2	0
DHML36--Lyl73/Vee"S"	118	106.4	4	8.3	18	41	92.4	0	0
DHML37--Lu26/3063	128	105.0	4	9.3	17	38	61.2	0	0
DHML41--Lu26/3063	120	97.8	5	7.7	15	41	62.4	0.2	0
DHML46--Lyl73/Vee"S"	123	103.4	6	7.3	19	50	152.4	0	0
DHML48--Lyl73/Vee"S"	128	123.2	4	6.3	14	33	60.0	0	0
DHML49--Lu26/3066	119	104.2	5	7.0	16	37	121.8	0	0
DHML50--Lu26/3062	118	87.8	5	8.7	17	63	123.6	0.2	0
DHML55--Lu26/3063	120	84.6	3	8.0	15	38	91.2	0	0
DHML56--C306/3013	117	109.2	3	7.7	17	38	61.2	0	0
Inquilab (check)	130	98.6	6	17.0	18	50	130.0	0.2	0

*Yr = yellow rust; Lr = leaf rust

others it was negatively correlated. This shows that higher concentration of proline in the genotypes under stress is not the sole indicator for drought selection.

Table 4. Days to maturity, grain yield and seed indices of DH-mutants under the rainfed conditions at different locations in North West Frontier Province

DH-mutant	Days to maturity			Grain yield (g/plot)			1000 grain weight (g)
	Tarnab	Karak	D.I.Khan	Tarnab	Karak	D.I.Khan	
DHML7	167 cd*	159 c	162 b	1000 cd	677 cd	750 a	37.4
DHML9	169 abc	160 c	158 c	1550 b	905 a	750 a	34.6
DHML12	165 d	161 bc	160 bc	1167 bcd	763 bc	602 bc	33.0
DHML13	170 ab	167 a	163 b	1100 cd	690 cd	607 bc	29.8
DHML20	169 bc	165 ab	159 bc	840 d	410 e	6507 bc	32.1
DHML27	162 e	152 d	157 c	1330 bc	695 cd	500 c	36.1
DHML46	168 bcd	159 c	160 bc	1217 bcd	618 d	488 c	35.0
DHML49	169 bc	161 bc	160 bc	1173 bcd	633 d	575 bc	29.8
DHML50	170 ab	165 ab	168 a	1693 a	840 a	760 a	29.7
Inqilab (check)	172 a	167 a	163 b	1123 bcd	614 d	535 c	42.2

*Numbers within columns followed by the same letter are not significantly different

Table 5. Proline accumulation in the glasshouse grown DH-mutants

DH-mutant	Proline (ug/g)
DHML7	41.25
DHML9	77.30
DHML12	75.55
DHML13	47.25
DHML20	59.40
DHML27	71.20
DHML46	38.70
DHML49	41.00
DHML50	70.30
Inqilab (check)	30.00

Conclusion

It can be concluded that recombination breeding followed by immediate application of gamma radiation and haploidy techniques (i.e. anther culture) not only created useful genetic variability for characters of economic importance but also reduced the breeding cycle time. Some of the desirable DH-mutants for higher yield, earliness and disease resistance, developed through our research endeavours are expected to lead directly to the development of improved wheat varieties for moisture stressed rainfed areas of Pakistan. It is mentioned that the selected DH-mutants will definitely need further zonal as well as countrywide field testing in the drought prone environments in order to confirm their adaptability and genetic stability.

Acknowledgement

The authors are gratefully acknowledge the financial help extended by the International Atomic Energy Agency (IAEA) for the partial funding of this research under Research Contract No.PAK-7644/RB. The authors are grateful to Director NIFA for constant encouragement. Great thanks are extended to Dr. Abid Farid, Mr. Saeed Gul and Mr. Sadaqat for statistical analysis and typing of the manuscript on a computer.

References

Chuang, C.C., T.W. Ouyang, H. Chia, S.M.M. Chou, and C.K. Ching, 1978. A set of potato media for wheat anther culture. In: Proc. Symp. Plant Tissue Culture. Science Press, Peking. pp.51-56

Ding, X.L., D.J. Luckett, and N.L. Darvey, 1991. Low dose gamma irradiation promotes wheat anther culture response. Aust. J. Bot. **39**: 467-474

Henry, Y., J. De-Buyser, M. Amssa, and G. Taleb, 1982. *In vitro* androgenesis in winter wheat breeding. In: Proc. 5th Intl. Cong. Plant Tissue and Cell Culture. Fujiwara, A. (Ed.) Abe Photo Printing Co., Tokyo. pp.569-570

Hu, D., 1996. A series of Jinghua winter wheat varieties developed by anther culture. In: Proc. 8th Assembly Wheat Breeding Society of Australia. pp.144-147

Hu, H., 1986. Wheat: improvement through anther culture. In: Biotechnology in Agriculture and Forestry 2. Bajaj, Y.P.S. (Ed.) Springer Verlag, Berlin. pp.55-72

Konzak, C.F. and H. Zhou, 1991. Anther culture methods for doubled haploid production in wheat. Proc. FAO/IAEA Meeting, Katowice, Maluszynski, M. and Z. Barabas (Eds.). Cereal Res. Commun. **19(1-2)**: 147-164

Kramer, P.J., 1980. Drought stress and the origin of adaptations. In: Adaptation of plants to water and high temperature stress. Turner, N.C. and P.J. Kramer (Eds.) John Wiley and Sons, New York. pp.7-20

Ouyang, J.W., 1986. Induction of pollen plants in *Triticum aestivum*. In: Haploids of Higher Plants *In Vitro*. Hu, H. and H. Yang (Eds.) China Academic Publishers, Springer-Verlag, Beijing,Berlin. pp.26-41

Ouyang, J.W., S.M. Zhou, and S.E. Jia, 1983. The response of anther culture to culture temperature in *Triticum aestivum*. Theor. Appl. Genet. **66**: 101-109

SAS Institute. (1990) SAS/STAT guide for personal computer, version 6.04. SAS Institute, Cary.

Singh, T.N., D Aspinall, L.G. Paleg, and S.F. Bogges, 1973a. Stress metabolism: II. Changes in proline concentration in excised plant tissues. Aust. J. Bio. Sci. **26**(57): 63

Singh, T.N., L.G. Paleg, and D Aspinall, 1973b. Stress metabolism: I. Nitrogen metabolism and growth in barley during water stress. Aust. J. Bio. Sci. **26**: 45-56

Stewart, C.R. and A.D. Hanson, 1980. Proline accumulation as a methabolic response to water stress. In: Adaptation of plants to water and high temperature stress. Turner, N.C. and P.J. Kramer (Eds.) John Wiley and Sons, New York. pp.173-189

Tan, B.H. and G.M. Halloran, 1980. Variation and correlations of proline accumulation in spring wheat cultivars. Crop Sci. **22**: 459-463

Zapata, F.J. and R.R. Aldemita, 1986. Induction of salt tolerance in high yielding rice varieties through mutagenesis and anther culture. In: Current Options for Cereal Improvement. Maluszynski, M. (Ed.) Kluwer Academic Publisher, Dordrecht. pp.193-202.

Anther culture in connection with induced mutations for rice improvement

Q.F. Chen, C.L. Wang, Y.M. Lu, M. Shen, R. Afza[1], M.V. Duren[1] and H. Brunner[1]
Zhejiang Academy of Agricultural Sciences, Hangzhou 310021, P.R.China
[1] *Plant Breeding Unit, FAO/IAEA Agriculture and Biotechnology Laboratory, Seibersdorf, Austria*

Key words: colchicines, doubled haploids, gamma rays, mutagenesis, mutants, *Oryza sativa*

Summary

Doubled haploids have long been recognized as a valuable tool in plant breeding since it not only offers the quickest method of advancing hetrozygous breeding lines to homozygosity, but also increases the selection efficiency over conventional procedures due to better discrimination between genotypes within any one generation. Ten cultivars of *japonica* rice and nine cultivars of *indica* rice were evaluated for androgenic response. Various doses (10 - 50 Gy) of gamma rays were applied to investigate the effect of radiation on callus formation, green plant regeneration and the frequency of selected doubled haploid mutants. Similarly, the effects of colchicine concentration (10-200 mg/L) on callus induction, regeneration and fertility of green plants were observed. It was demonstrated that the dose of 20 Gy gamma rays and 30 mg/L concentration of colchicine have significant stimulation effect on regeneration of green plants from rice anther culture. The high frequency of observed doubled haploid mutants indicates that anther cultures applied in connection with gamma rays is an effective way to improve rice cultivars.

M. Maluszynski and K.J. Kasha (eds.),
Mutations, In Vitro and Molecular Techniques for Environmentally Sustainable Crop Improvement, 83–92.

Introduction

Haploid development by *in vitro* methods includes anther/pollen culture, ovary culture and chromosome elimination. Among them, anther culture is one of the simplest, most efficient and most practical techniques. Doubled haploids (DH) have long been recognized as a valuable tool in plant breeding since it not only offers the quickest method of advancing hetrozygous breeding lines to homozygosity, but also increases the selection efficiency over conventional procedures due to better discrimination between genotypes within any one generation (Croughan *et al.*, 1987; Veilleux, 1994; Martinez *et al.*, 1996; Yang, 1997; Chen *et al.*, 1997). In addition, anther culture could also be an effective vehicle for producing variation as it allows early expression of recessive genes. Although anther culture has already been applied in plant breeding and some new cultivars have been successfully released by the technique, its application to the improvement of crops has been hampered by the difficulty of inducing morphogenesis. The frequency of DH plant regeneration is not always high and predictable for different crops and cultivars (Keshari *et al.*, 1992; Yan *et al.*, 1996; Raina and Zapata, 1997). This paper reports on high efficiency of doubled haploid production through anther culture and their applications in combination with mutation techniques for rice improvement.

Materials and methods

Culture media

Ten cultivars of japonica rice and nine cultivars of indica rice were evaluated for androgenic response in anther culture. For anther culture the induction medium was the N_6 basal medium supplemented with 10 μM 2,4-D 60, g/l sucrose and 1.8 g/l gelrite, pH 5.8. Regeneration medium was the MS basal medium supplemented with 9.0 μM BAP, 2.7 μM NAA, 100 mg/L myo-inositol, 100 mg/L casein hydrolysate, 200 mg/L thiamine, 30 g/l sucrose, 9g/l agar, pH 5.8.

Methods

Rice boots at mid and late-uninucleate stage of pollen development were collected respectively and subjected to cold pretreatment at 7°C for 5 days. The panicles were surface disinfected by immersing in 70% ethanol for 20 seconds and in 50% commercial Clorox or 0.5% sodium hypochloride solution for 10 minutes, then washed with sterile distilled water 3 times. Anthers were cultured on semisolid induction medium in 2.5/15 cm tubes (50 anthers each, in 10 tubes per treatment) and incubated at 27°C under dark conditions. Gamma rays were applied after 30-35 days of incubation.

Calli of about 1-2 mm in diameter derived from the anthers after irradiation were transferred to the regeneration medium in 3.0 cm×20 cm tubes, 2 calli per tube, and then incubated at 27°C under constant light for about 40 days. Transfer of calli subsequently emerging from anthers was done at 10 day intervals. Green plantlets were cultured and subcultured on MS basal medium supplemented with 5μM NAA until the shoot tip reached the top of the tubes, then the tubes were opened and moved from the growth room to the greenhouse and kept shaded for 2-3 days. A 1-2 cm layer of distilled water was then added above the medium and tubes were moved from shade into light for 2-3 days. The plantlets, after 5 days acclimatization were transplanted to pots with soil or to the field. The ploidy level of the plants was examined by flow cytometry and estimated by observation of morphological characters. For chromosome doubling, haploid plants were treated by immersing roots into 1.25 mM colchicine or 25 μM oryzalin for 12 h at tillering stage.

To investigate the effect of radiation on callus formation and green plant regeneration 10, 20, 30, 40 and 50 Gy gamma rays (^{60}Co) were used. The frequency and selection of doubled haploid mutants (MDH) was performed in progenies obtained from anther culture irradiated with 10, 20 and 30 Gy.

Results and discussion

Improvement of anther culture procedures

The best concentration of 2,4-D for the formation of calli was 10-15 μM (Fig.1). At higher or lower concentrations, the number of calli derived from anthers decreased. Boots cut from rice plants were wrapped in aluminum paper, then put in a cold room at 7°C for different numbers of days before culture. The results showed that the number of calli increased with the increase of pretreatment time and reached the highest number on the fifth day (about 0.5 times higher than that of non-treated control), and then decreased rapidly with length of time (Fig.2). The ability to form calli was almost lost after 11 days.

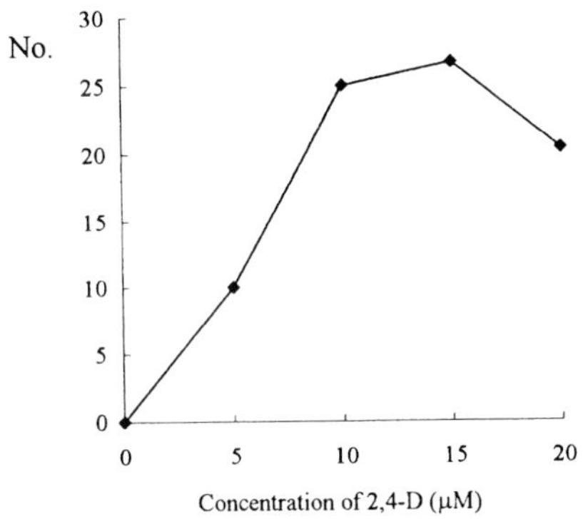

Figure 1. Relationship between the number of calli formed (No./100 anthers) and concentration of 2,4-D supplemented in the induction medium (cv. Taipei 309)

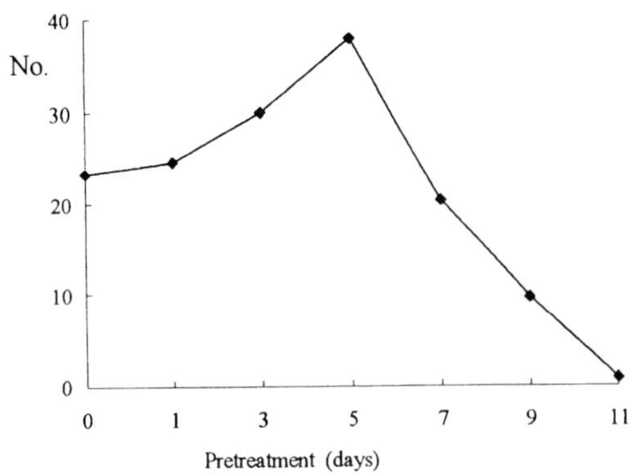

Figure 2. Relationship between the number of callus formation (No./100 anthers) and duration of anther pretreatment at 7°C before culture (cv. Taipei 309)

The formation of calli and regeneration of green plants from the anthers at mid-uninucleate and late-uninucleate stages were investigated. The results showed that the anthers at mid-uninucleate stage promote callus formation (Table 1). The time of callus initiation and number of calli formed from the anthers were 10 days earlier and 36.6% higher than from anthers at the late-uninucleate stage, but the anthers at late-uninucleate stage favored differentiation and regeneration of green plants. The number of calli producing green plantlets and regenerated green plants were 3.1 and 4.1 times higher than from anthers at the mid-uninucleate stage, respectively.

Table 1. Effects of anther stage on the formation of calli and regeneration of green plants in rice cv. Taipei 309

Anther stage	Number of anthers cultured	Time of callus initiation (days)	Formation of calli		Calli with green plantlets		Regeneration of green plants	
			(No.)	(%*)	(No.)	(%*)	(No.)	(No./100 anthers)
mid-uninucleate	500	25	340	68.0	23	6.8	481	96.2
late-uninucleate	500	35	250	50.0	70	28.0	2459	491.8

* Total number of calli transferred/total number of anthers cultured ×100
** Number of calli with green plantlets/total calli transferred to the regeneration medium

Two thousand anthers collected at late-uninucleate microspore stage were cultured on the induction medium. Transfer of first group calli to the regeneration medium was on the 35th day of incubation. The transfer of calli emerging during later periods from anthers was done at 10 days intervals. The results showed that the number of calli with green plantlets increased with the time of anther incubation (35 days - 7.7%; 45 days - 10.6%; 55 days - 16.7%), but the differentiation ability of green plantlets was lost for all calli after 65 days. The differentiation of green spots was not very different among the different times of anther incubation, except being lower after 65 days. The calli with highest ability of green plant regeneration were those transferred at 35 days of incubation and differentiating green plantlets at about the 25th day after transfer to regeneration medium. The regeneration of green plants reached 443.6 plants per 100-anthers cultured and was much higher than that of the calli transferred and differentiating green plantlets later on (Table 2).

Table 2. Relationship between regeneration of plants and the time of anther incubation in rice cv. Taipei 309

Anther incubation (days)	Calli transferred (No.)	Calli with green plantlets (No.)	(%)	Calli with green spots (No.)	(%)	Calli with albino plants (No.)	(%)	Regeneration of green plants (No.)	(No./100 anthers)
35	312	24	7.7	49	15.7	109	34.9	2218	443.6
45	404	43	10.6	66	16.3	109	34.9	103	20.6
55	450	75	16.7	75	16.7	113	25.1	524	104.6
65	272	0	0.0	18	6.6	28	10.3	0	0.0

Eighteen rice cultivars were used to determinate the ability to form calli and regenerate green plants from non-irradiated anther cultures. The results showed that of the 9 cultivars of *japonica* rice, 8 cultivars had the ability for callus formation, but only 3 cultivars regenerated green plants under the described procedure (Table 3). Among them, the best cv. was 'Wang Shi bo'. From 9 cultivars of *indica* rice, only 3 cultivars formed calli and 1 cultivar regenerated green plants. The problem of different anther culture ability of *japonica* and *indica* rice was also reported by other authors (Zapata-Arias *et al.*, 1995; Yan *et al.*, 1996; Tang *et al.*, 1998; Yamagishi *et al.*, 1998; Faruque *et al.*, 1998).

Table 3. The ability for callus formation and green plant regeneration of various rice cultivars

Cultivar	Formation of calli		Calli with green plantlets		Green plants regenerated	
	(No.)	(%)	(No.)	(%)	(No.)	(No./ 100 anthers)
japonica rice						
Wang shi bo	152	30.4	5	3.3	24	4.8
Zao jing 26	148	29.6	2	1.4	7	1.4
Zhe jing 26	139	27.8	2	1.4	5	1.0
Zao jing 12	131	26.2	0	0.0	0	0.0
Zao jing 19	126	25.2	0	0.0	0	0.0
Liao jing 10	102	20.4	0	0.0	0	0.0
Yue guang	89	17.8	0	0.0	0	0.0
Zao jing 28	45	9.0	0	0.0	0	0.0
Zao jing 15	0	0.0	0	0.0	0	0.0
indica rice						
Zhong shi 2	127	25.4	1	0.8	3	0.6
Zhou you 6	116	23.2	0	0.0	0	0.0
Zhong you zao 2	39	7.8	0	0.0	0	0.0
Jia xing 13	0	0.0	0	0.0	0	0.0
Zhe 9248	0	0.0	0	0.0	0	0.0
Hang 93-1	0	0.0	0	0.0	0	0.0
Zhou 903	0	0.0	0	0.0	0	0.0
G93-368	0	0.0	0	0.0	0	0.0
Jia xing 35	0	0.0	0	0.0	0	0.0

Stimulatory effect of gamma rays and colchicine on the formation of calli and green plant regeneration

Almost all doses, from 10 to 50 Gy of gamma rays, favored the formation of calli and the regeneration of green plants (Table 4). The best dose for the formation of calli was 10 Gy where the number of calli formed was 81.7% higher than that of the non-irradiated control. For green plant regeneration, the best dose was 20 Gy where the number of green plantlets and regenerated green plants per 100 anthers cultured was 491, which was more than 300 times higher than that of non-irradiated control, respectively. The differentiation of green spots was higher and albino plant formation was lower at all doses of gamma rays than that of the non-irradiated control (Lu et al., 1999).

The number of calli producing green plantlets and regenerated green plants from the anthers of 6 cultivars irradiated with 20 Gy of gamma rays was 11.9-18.6%. This was 2-4 times and 12-38 times higher than that of non-irradiated controls, respectively (Table 5).

The calli from non-irradiated anthers were inoculated on the regeneration media containing different concentrations of colchicine and the effects of colchicine concentration on differentiation, regeneration and fertility of green plants were observed (Table 6). The results indicated that low concentrations (up to 50 mg/L) of colchicine improved the differentiation

Table 4: Effect of gamma rays dose on the formation of calli and regeneration of green plants in rice cv. Taipei 309

Dose (Gy)	Induction of calli		Calli with green plantlets		Calli with green spots		Calli with albino plants		Regeneration of green plants	
	(No.)	(%)	(No.)	(%)	(No.)	(%)	(No.)	(%)	(No.)	(No./100 anthers)
0	164	32.8	2	1.2	16	9.8	62	37.8	8	1.6
10	298	59.6	27	9.1	45	15.1	70	23.5	214	42.8
20	250	50.0	70	28.0	26	10.4	35	14.0	2459	491.8
30	230	46.0	32	13.9	35	15.2	64	27.8	80	16.0
40	252	50.4	2	0.8	44	17.5	61	24.2	6	1.2
50	244	48.8	9	3.7	42	17.2	67	27.5	78	15.6

Table 5. Effect of gamma rays on callus formation and green plant regeneration of different rice cultivars in anther culture

Cultivar	No. of anthers cultured	Dose of gamma rays (Gy)	Formation of calli		Calli with green plantlets		Regeneration of green plants	
			(No.)	(%)	(No.)	(%)	(No.)	(%)
japonica rice								
Wang shi bo	500	0	161	-	6	-	28	-
	500	20	191	+18.1	31	+417.0	1056	+3671.9
Zao jing 26	500	0	143	-	3	-	9	-
	500	20	160	+11.9	11	+267.0	127	+1311.0
Zhe jing 26	500	0	129	-	2	-	5	-
	500	20	145	+12.4	7	+250.0	81	+1520.0
indica rice								
Zhong shi 2	500	0	120	-	1	-	4	-
	500	20	135	+12.5	5	+400.0	52	+1200.0
Zhou you 6	500	0	111	-	0	-	0	-
	500	20	125	+12.6	3	+	19	+
Zhong you zao 2	500	0	45	-	0	-	0	-
	500	20	52	+15.6	2	+	7	+

and regeneration of green plants from calli, but higher concentrations (more than 75 mg/L) were obviously harmful to inoculated calli and the regeneration of plants. When the concentration was more than 100mg/L, almost all inoculated calli and regenerated plants died. The survival rate of regenerated green plants decreased with an increase of colchicine concentration in the media, but the rate of fertile green plants increased with an increased concentration. 20-50 mg/L was better concentration for differentiation, regeneration and fertility of green plants from rice anther culture (Table 6). In addition, the regenerated green plants had to be transferred to MS basal medium without colchicine, otherwise the plants would die. Increased doubled haploid plant regeneration from rice anthers culture on colchicine supplemented media was also reported by Alemanno and Guiderdoni, 1994.

Table 6. Effect of colchicine concentration in regeneration medium on green plant regeneration

Cultivar	Concentration of colchicine (mg/L)	Inoculated calli (No.)	Calli differentiating green plants (No.)	(%)	Green plants regenerated (No.)	Mean No. of green plants/ 100 calli	Surviving plants (No.)	(%)	Fertile plants (No.)	(%)
Zao jing 26	0	150	2	1.33	11	7.33	11	100.00	1	9.09
	10	142	3	2.11	39	27.46	39	100.00	6	15.38
	20	162	5	3.09	78	48.15	71	91.03	18	25.35
	30	150	6	4.00	82	54.67	74	90.24	21	28.38
	40	124	5	4.03	54	43.55	45	83.33	13	28.89
	50	166	7	4.22	68	40.96	55	80.88	18	32.37
	75	114	3	2.63	21	18.42	5	23.81	2	40.00
	100	158	2	1.27	5	3.16	0	0.00	0	0.00
	200	148	0	0.00	0	0.00	0	0.00	0	0.00
Zhou you 6	0	132	1	0.75	7	5.30	7	100.00	1	14.29
	10	140	2	1.43	18	12.86	17	94.44	3	17.65
	30	142	4	2.82	45	31.69	40	88.89	9	22.50
	50	140	4	2.86	38	27.14	30	78.95	8	26.67
	75	132	2	1.52	12	9.09	2	16.67	1	50.00
	100	130	1	0.77	2	1.54	0	0.00	0	0.00

Twenty Gy of gamma rays and 30 mg/L of colchicine had obvious stimulatory effects on green plant regeneration from rice anther culture. The stimulation effect of gamma rays combined with colchicine was much better than that of either used alone (Table 7).

Table 7. The stimulation effect of gamma rays combined with colchicine on green plant regeneration

Cultivar	Treatments		Differentiation of green plants				Regeneration of green plants				
	γ-rays (Gy)	colchicine (mg/L)	calli inoculated (No.)	calli with green plants (No.)	(%)	total (No.)	mean per 100 calli (No.)	survival (No.)	(%)	fertile (No.)	(%)
Zao jing 26	0	0	150	2	1.33	11	7.33	11	100.00	1	9.09
	0	30	150	6	4.00	82	54.67	74	90.24	21	28.38
	20	0	164	9	5.49	118	71.95	113	95.76	14	12.39
	20	30	160	12	7.50	156	97.50	140	89.74	30	21.43
Zhou you 6	0	0	132	1	0.76	7	5.30	7	100.00	1	14.29
	0	30	142	4	2.82	45	31.69	40	88.89	9	22.50
	20	0	136	5	3.68	67	49.26	61	91.04	10	16.39
	20	30	136	7	5.15	79	58.09	68	86.08	16	23.53

Application of anther culture in rice improvement through induced mutations

In total, 3,369 green plants were regenerated from irradiated anther culture of rice cultivar Taipei 309. Among them 70.7% were haploid, 19.3% diploid and the others were triploid, tetraploid and mixploid. While 31.0% of haploid plants treated with 25 μM oryzalin were doubled, only 11.5% of haploid plants treated with 1.25 mM colchicine were doubled. In total 636 plants were chromosone doubled and fertile. Some variations were found among regenerated doubled haploids (MDH1). These included early and late maturity, short and tall stem, plant shape and habit, chlorophyll offtypes (striata), panicle and grain size.

The total mutation frequency in the next generation (MDH2) ranged from 31.8 (10 Gy) to 67.7% (30 Gy) in the treated combinations in comparison to 32.8% in the nonirradiated control. From the entire experiment 90 early maturing, 50 late maturing and 59 short stem mutants were selected. Two early maturing and 1 semidwarf DH lines were obtained from nonirradiated combinations. The mutations of deformity included change of plant shape, leaf shape, chlorophyll deficiency and sterility. Among 67 DH lines from the non-irradiated control all 19 DH lines with deformity were sterile, tetraploidy and had awns (Table 8). The successful application of induced mutations and anther culture for improvement of agronomically important characters was also reported by other authors (Zapata and Aldemita, 1986; Aldemita and Zapata, 1991; Chen *et al.*, 1996)

Table 8. Frequency of mutations in MDH2 generation

Doses (Gy)	Plant lines	Mutated character									
		Heading date		Plant height		Characters related to high yield		Deformity		Total	
	(No.)	(No.)	(%)	(No.)	(%)	(No.)	(%)	(No.)	(%)	(No.)	(%)
0	67	2	3.0	1	1.5	0	0.0	19	28.4	22	32.8
10	220	51	23.2	8	3.6	4	1.8	7	3.2	70	31.8
20	188	48	25.5	8	4.3	10	5.3	11	5.8	77	41.0
30	161	41	25.5	43	26.7	8	5.0	17	10.6	109	67.7

In total, 40 (6.3%) DH mutant lines with agronomically important characters were selected for further observation from a total of 636 lines in the MDH2. Among them, were 8 lines with mutations for earlier maturity, (one of them was 2 weeks earlier than parent cultivar); 10 lines with shorter stem and 22 lines with improvement in characters related to high yield. Four DH mutant lines with promising characteristic have been selected in MDH3 for further agronomic evaluation and named Atom-1 (Am-1), Atom-3 (Am-3), Atom-11 (Am-11) and Atom-12 (Am-12). Am-1 and Am-3 were used as early season rice and planted at the north, middle and south regions of Zhejiang Province. The main agronomic characters of Am-1, Am-3 and check cultivar 'Zhe 733' were listed in Table 9.

Table 9. Mean value of main agronomic characters of Am-1, Am-3 and check cultivar Zhe 733

Lines	Growing period (days)	Valid tillers (No.)	Plant height (cm)	Length of ear (cm)	Total grains per ear (No.)	Setting percentage (%)	Thousand grain weight (g)
Am-1	115	3.0	93.1	19.8	125.0	83.9	22.8
Am-3	128	3.7	100.8	20.0	157.5	68.5*	28.4
Zhe 733 (Check)	115	2.4	78.7	19.1	97.1	72.5	26.2

* just met the vile weather during heading period

The result indicates that Am-1 is a promising line since its agronomic characters are much better than that of the check cultivar, except for thousand grain weight. It was recommended for regional testing in 1998. Am-3 can only be used for single-cropping rice since its growing period is too long to be used for early rice. Neither of them was infected rice blast when tested in a special pathogenetic region. The grain quality of Am-1 and Am-3 reached the standard of second grade rice, which was issued by Chinese Agricultural Ministry (NY 122-86). Protein content of them was up to 11.8% and 11.7% respectively, which was 25.5% and 24.5% higher than that of check cultivar Zhe 733 (9.4%).

Am-11 and Am-12 were used as late season rice and planted in the experimental field at Hangzhou (north Zhejiang Province). The mean plot yield of Am-11 and Am-12 was 6.12% and 2.04% higher than that of check cultivar 'Xiushui 11' respectively, but only Am-11 reached 5% significance level. The main agronomic characters and disease-resistance of Am-11 were much better than that of the check cultivar Xiushui 11 except for lodging resistance.

Conclusions

- A dose of 20 Gy gamma rays and 30 mg/L concentration of colchicine have an obvious stimulation effect on regeneration of green plants from rice anther culture.
- The technique of anther culture applied in connection with induced mutations is an effective way to improve rice cultivars. The mutation frequency and selection efficiency were greatly increased, and subsequent generation of DH mutants was stable.
- One line, Am-1, has been recommended for regional trials.

References

Aldemita, R.R. and F.J. Zapata, 1991. Anther culture of rice: effects of radiation and media components on callus induction and plant regeneration. Cereal Res. Commun. **19**: 9-32

Alemanno, L. and E. Guiderdoni, 1994. Increased doubled haploid plant regeneration from rice (*Oryza sativa* L.) anthers cultured on colchicine-supplemented media. Plant Cell Rep. **13**: 432-436

Chen, Q.F., R. Afza, M. van Duren, and H. Brunner, 1997. Effects of gamma irradiation and other factors on the induction of calli and regeneration of green plants in rice anther culture. Acta Agri. Nucl. Sinica. **11**(1): 59-64

Chen, Q.F., C.L. Wang, Y.M. Lu, and M. Shen, 1996. Rice induced mutation breeding by irradiated anther culture. Acta Agric. Zhejiangensis. **8**(4): 202-207

Croughan, T.P., Q.R. Chu, K.S. McKenzie, H.J. Huang, and M.M. Meche, 1987. The use of anther culture to expedite the breeding and release of new varieties of rice. Annu. Prog. Rep. La. Agric. Exp. Stn. pp.52-52

Faruque, M.O., T. Farzana, Z.I. Seraj, R.H. Sarker, and A.A. Khatun, 1998. Variation in green plant regeneration response from anthers of *indica* rice and their hybrids with *japonica* cv. Taipei 309. Plant Cell Tiss. Org. Cult. **54**: 191-195

Keshari, A.K., R.L. Singh, S.K. Agrawal, G.K. Agrawal, and V.P. Agrawal, 1992. Studies on anther culture of rice cultivars of Nepal. Biotech. Biodiv. Lett. **2**: 1-10

Lu, Y.M., C.L. Wang, M. Shen, and Q.F. Chen, 1999. Effect of γ-irradiation on the formation of calli and regeneration of green plants in rice anther culture. Acta Agric. Zhejiangensis. **9**(3): 123-216

Martinez, C.P., F. Correa Victoria, M.C. Amezquita, E. Tulande, G. Lema, and R.S. Zeigler, 1996. Comparison of rice lines derived through anther culture and the pedigree method in relation to blast (*Pyricularia grisea* Sacc.) resistance. Theor. Appl. Genet. **92**: 583-590

Raina, S.K. and F.J. Zapata, 1997. Enhanced anther culture efficiency of *indica* rice (*Oryza sativa* L.) through modification of the culture media. Plant Breed. **116**: 305-315

Tang, K., X. Sun, Y. He, and Z. Zhang, 1998. Anther culture response of wild *Oryza* species. Plant Breed. **117**: 443-446

Veilleux, R.E., 1994. Development of new cultivars via anther culture. HortSci. **29**(11): 1238-1241

Yamagishi, M., M. Otani, M. Higashi, Y. Fukuta, K. Fukui, M. Yano, and T. Shimada, 1998. Chromosomal regions controlling anther culturability in rice (*Oryza sativa* L.). Euphytica **103**: 227-234

Yan, J., Q. Xue, and J. Zhu, 1996. Genetic studies of anther culture ability in rice (*Oryza sativa*). Plant Cell Tiss. Org. Cult. **45**: 253-258

Yang, Z., 1997. Inheritance of photoperiod sensitive genetic male sterility and breeding of photoperiod sensitive genic male-sterile lines in rice (*Oryza sativa* L.) through anther culture. Euphytica **94**: 93-99

Zapata-Arias, F.J., L.B. Torrizo, and A. Ando, 1995. Current developments in plant biotechnology for genetic improvement: the case of rice (*Oryza sativa* L.). World J. Microbiol. Biotech. **11**: 393-399

Zapata, F.J. and R.R. Aldemita, 1986. Induction of salt tolerance in high yielding rice varieties through mutagenesis and anther culture. In: Current Options for Cereal Improvement. Maluszynski, M. (Ed.) Kluwer Academic Publishers, Dordrecht. pp.193-202.

Production of salt tolerant rice mutants using gamma rays and anther culture

W. Navarro Alvarez
Escuela de Ciencias Agrarias, Universidad Nacional, Heredia, Costa Rica

Key words: Oryza sativa, salinity tolerance, doubled haploids, mutants, earliness

Summary

Rice is known to be less resistant to salt than many other crops. Excess of salt in soil limits the yield potential of most rice varieties. New rice genotypes with higher adaptation to salinity are therefore needed. In this work the combination of mutation induction and anther culture were used to produce rice mutants with tolerance to salinity. Seven commercial varieties of Costa Rica were irradiated with gamma rays from a Cobalt-60 source. Sixty four mutants from treated populations were selected as resistant to salinity. The number of productive tillers, number of grains per panicle and panicle length were much higher in some mutants than in parent varieties. It was also possible to select mutants with increased number of days to heading, a critical factor that limited production in parent varieties.

M. Maluszynski and K.J. Kasha (eds.),
Mutations, In Vitro and Molecular Techniques for Environmentally Sustainable Crop Improvement, 93–96.

Introduction

Soil salinity is one of the major constraints for rice production in developing countries. The problem is more acute in arid and semi-arid regions where irrigation with brackish water, poor drainage, and high evaporation combine to concentrate salts in root zones of crop plants.

Under field cultivation, rice is known to be less resistant to salinity than some other crops, such as barley (Epstein *et al.*, 1980). Excess of salt in the soil limits the yield potential of most rice varieties. The concentration as low as 50 mM of NaCl is already lethal at the seedling stage (Yeo *et al.*, 1990). According to Tadano (1983), causes of injury to crop growth by high concentrations of salt include: the inhibition of water absorption by an increase of osmotic pressure in the soil; growth injury by an increase in salt content in the plant; specific growth injury by elements consisting of high concentrations of salts involved also in indirect growth injuries due to the inhibition of absorption of essential nutrients. Important components of grain yield such as the number of productive tillers, number of grains per panicle and panicle length are affected and reduce grain yield. Also, other morphological and physiological traits such as plant height and days to heading can be affected at high salt concentrations.

In order to produce new rice genotypes with better adaptation to salinity of the Costa Rican commercial rice cultivars, mutation techniques in combination with anther culture were used to develop desired mutants.

Materials and methods

Seeds of the seven commercial varieties of Costa Rica ('CR 1821', 'CR 1113', 'CR 5272', 'SETESA 9, 'CR 750', 'WS', 'CR 9027' and 'CAMAGO 8') were irradiated with gamma rays from a Cobalt-60 source, using a dose of 250 Gy. All irradiated seeds were sown in the field and grown to produce the M_1 and M_2 seed generations. When the M_1 plants had immature tillers, some of them were collected for anther culture purposes. The tillers were kept in the refrigerator for 5 days and then surface disinfected for 20 minutes with commercial bleach solution (20% bleach; 80% distilled water) with periodic agitation. After surface disinfection of the tillers, anther were aseptically excised and placed on the surface of the initiation medium FJ-4 for embryo induction. When embryo-like structures were observed, they were transferred to the FJ-4 regeneration medium modified with 1.25% NaCl for plant selection. Plants developed under these *in vitro* conditions, were transplanted to pots with soil and grown to maturity in greenhouse. Seeds were harvested individually from each plant. Progeny of each doubled haploid plant was evaluated in the field under saline conditions.

For the second part of the experiment, the M_1 plants raised after mutagenic treatment were harvested by taking one spike per plant. All harvested seed were mixed (bulked) and used to obtain the M_2 plant generation. Effective greenhouse procedures to screen for salinity tolerance have been developed at IRRI and CIAT. The following is the screening procedure used to test salinity tolerance for both, the doubled haploid plants produced by anther culture and segregating M_2 plants produced directly from M_1 plants.

Soil prepared with ground rice straw and dry soil, was added to plastic trays. Sea water was added at a ratio of 4:1 for each 5 kilos of soil. With this amount of sea water and soil, an EC (electrical conductivity) of 8-10 dS/m was achieved after the soil was dried. This is the upper limit of tolerance of rice plants to saline conditions. Therefore, plants growing under these conditions should be tolerant to salinity. Soil salt concentration was tested after each selection by the Soil Science Laboratory of the Universidad Nacional.

The M_2 and the parent variety seeds were sown in saline soil. After six weeks of growth under these conditions, the plants with good performance were transplanted to a rice field, with good climate conditions to enable sufficient seed production. All plants were harvested individually and kept separately for progeny evaluation.

For evaluation, each progeny was sown in a progeny/row design, using 4 m rows. All the seeds produced by each mutant plant were planted. After every ten mutant progenies, one row of a related parental variety was included as a check. The level of soil salinity was evaluated to determine the salinity conditions in the field at the beginning and at the end of the experiment.

Results and discussion

In vitro selection

The presence of NaCl (1.25%) in the regeneration medium was effective as a selection agent in the experiment. Sixteen doubled haploid plants were selected as salinity tolerant from the seven commercial varieties cultured. An additional effect was that too many albino plants were regenerated on the saline medium and some abnormal plants were also observed.

Induced mutants in M_2 and M_3

From about 10,000 M_2 plants planted on saline soil 64 mutants were selected in M_3 generation and identified as salt tolerant. The seed germination as well as the survival percentages of seedling were reduced by salinity in mutated generations of all varieties. Mutated populations from various varieties differed considerably in their responses to salinity. Root elongation of salt sensitive seedlings was considerably affected by salinity during the first week after germination, but not of salt resistant mutants.

At reproductive stage, the percentage of flowering plants was reduced and the flowering was delayed in sensitive plants under stress conditions. Different responses were observed among the resistant mutants. Some of them showed earliness while others expressed normal to late heading. The number of productive tillers, number of grains per panicle and panicle length, were shown also to differ in the response to stress (Table 1). In non-irradiated plants these traits decreased for all varieties while mutants showed a wide range of variability. The general appearance was also different between mutants and parent varieties under salinity conditions, especially the leaf colour became lighter in plants of parental varieties. Mutants with reduced plant height were selected in M_3 generation of all varieties.

The results showed that the most important traits related to the yield were affected by salinity under presented field conditions. Similar results were observed by Brinkman (1982) and Rana (1985). They found that salt stress induced a wide range of flower morphological modifications leading to decrease of floret fertility, the trait that is important yield component. In the contrast, these characters were not affected in some of the resistant mutants in presented experiment. For all of these characteristics, selected mutant lines were superior to their parental varieties under salinity conditions. Even thought several authors have suggested that salinity responsiveness is likely to be controlled by a number of genes (Foolad and Jones, 1991; Asins *et al.*, 1993), the presented results demonstrated that induced mutations can be used as a method to improve resistance to salinity in rice susceptible varieties.

Table 1. Range values of some agronomic characters of salinity tolerant mutant lines

Character	Range value of parent varieties	Range value of mutant lines
Number of productive tillers	18-23	29-65
Number of grains per panicle	78-119	81-205
Panicle length (cm)	17.7-23.5	17.2-28.7
Plant height (cm)	71-84	67-94
Days to heading	118-123	80-131

References

Asins, M.J., M.P. Breto, M. Cambra, and E.A. Carbonell, 1993. Salt tolerance in *Lycopersicon* species. I. Character definition and changes in gene expression. Theor. Appl. Genet. **86**: 737-743

Brinkman, R., 1982. Rapid reclamation of brackish water freshponds in acid sulfate soils. In: Proc. Symp. on Acid Sulfate Soils. The Netherlands. pp.318-320

Epstein, E., J.D. Norlyn, D.W. Rush, R.W. Kingsbury, and D.B. Kelley, 1980. Saline culture of crops: a genetic approach. Science. **210**: 399-404

Foolad, M.R. and R.A. Jones, 1991. Genetic analysis of salt tolerance during germination in *Lycopersicon*. Theor. Appl. Genet. **7**: 321-326

Rana, R.S., 1985. Breeding for salt resistance. Concept and strategy. Int. J. Tropical Agric.

Tadano, T., 1983. Salt tolerance in crop plants and its mechanism. In: Science of the Rice. Vol. 2. Physiology. Food and Agriculture Research Center, Tokyo. pp.983-991

Yeo, A.R., M.E. Yeo, T.J. Flowers, and S.A. Flowers, 1990. Screening of rice (*Oryza sativa* L.) genotypes for physiological characters contributing to salinity resistance and their relationship to overall performance. Theor. Appl. Genet. **79**: 377-384.

Induction of photoperiod sensitive genetic male steriles for use in hybrid rice seed production

J.N. Rutger
Dale Bumpers National Rice Research Center, USDA-ARS, P.O. Box 287, Stuttgart, Arkansas 72160 USA

Key words: Oryza sativa, genetic male sterility, induced mutations, gamma rays

Summary

The most notable photoperiod sensitive genetic male sterile in rice is the Nongken 58S line in China. Two sources of environmentally sensitive genetic male sterile mutants were found in California. The photoperiod-sensitive genetic male steriles (PGMS) have the potential for reducing hybrid rice breeding from the present 3-line system to a 2-line system. The long-range objectives of the present project were to induce an array of genetic male steriles in rice, and then to evaluate these at different day lengths in order to find steriles that become fertile in selected environments. Three genetic male steriles were selected from gamma irradiation of southern US cultivars. Inheritance of these mutants was studied through progeny tests in M_5 to M_7 generations and through fertility characterization of pollen and spikelets. Seed set on steriles under open pollination ranged from a low of 8.1% for Cypress 1819 to a high of 32.9% for Kaybonnet 1789.

M. Maluszynski and K.J. Kasha (eds.),
Mutations, In Vitro and Molecular Techniques for Environmentally Sustainable Crop Improvement, 97–100.

Introduction

The long-range objectives of the present project were to induce an array of genetic male steriles in rice, and then to evaluate these at different day lengths in order to find steriles that become fertile in selected environments. Such photoperiod-sensitive genetic male steriles (PGMS) have the potential for reducing hybrid rice breeding from the present 3-line system to a 2-line system.

The most notable photoperiod sensitive genetic male sterile in rice is the Nongken 58S line in China (Shi, 1985), which is reported to be sterile at day lengths of 14 hours or more, and fertile at photoperiods less than 13 hours and 45 minutes (Yuan *et al.*, 1993). Two sources of environmentally sensitive genetic male sterile mutants were found in California (Oard *et al.*, 1991; Rutger and Schaeffer, 1994). Since the fertility on/off switching processes in the two California sources are still not fully understood, it would be highly desirable to have additional environmentally sensitive male steriles, preferably one in which sterility is sensitive only to photoperiod. Four experiments to identify such additional environmentally sensitive male steriles were conducted.

Materials and methods

Experiment 1

In the summer of 1994, 511 spontaneously occurring sterile plants were ratooned to the greenhouse from farm fields. Day length in the farm fields was about 14 hours. In the greenhouse the ratoons were kept at natural day lengths, of 12 hours or less. Plants reverting to fertility were progeny tested in summer 1995.

Experiment 2

In the summer of 1995, a total of 141 M_2 lines, from seven cultivars, segregated for sterility. Progenies of eight fertile sibs within each M_2 family were sent to the 1995/96 Puerto Rico winter nursery, which has day length of 12 hours or less.

Experiment 3

In the summer of 1997, 56 M_2 lines, from a 200 Gy gamma ray-treated population of about 1,000 M_2 lines of the cultivar 'Kaybonnet', segregated for sterility. Progeny tests of four fertile sibs within each segregating M_2 family were sent to the 1997/98 Puerto Rico winter nursery.

Experiment 4

During the course of searching for PGMS types in Experiment 2, the inheritance of three non-PGMS steriles was studied.

Results and discussions

Experiment 1

Sixty of the 511 spontaneous steriles reverted to varying degrees of fertility in the winter greenhouse. These sixty ratoons were transplanted to the field in summer 1995. Also, they were progeny tested by sowing seed collected during the winter. All 60 ratoons and their progenies were fertile in summer 1995. Had any of these been PGMS types, they should have become sterile again in 1995. It was concluded that PGMS was not present in these materials.

Experiment 2

In the summer of 1995 M_2 populations, the 141 M_2 lines that segregated for within-in line sterility were judged to be appropriate sources of steriles. In the absence of PGMS reactions, two-thirds of the eight fertile sibs of each line should segregate for sterility in the Puerto Rico winter nursery. Failure to segregate for sterility is the desired indication of PGMS. Such techniques were keys for discovering the two environmentally sensitive male steriles previously found in California (Oard *et al.*, 1991; Rutger and Schaeffer, 1994), and were expected to work here as well.

In the 1995/96 winter nursery these 141 lines were narrowed down to 41 which failed to segregate for sterility. In the 1996 summer nursery, these 41 were further narrowed down to just one line, which produced a progeny row of sterile plants. This would be expected from a msms plant which behaved as fertile in the winter (short-day) nursery. However, in subsequent generations this line began behaving as a regular recessive male sterile, i.e., not PGMS.

Experiment 3

In the 1997/98 winter nursery the 56 lines were narrowed down to 5 which failed to segregate for sterility. These were progeny tested in the 1998 summer nursery at Stuttgart. None proved to be the PGMS type.

Experiment 4

Three genetic male steriles were selected from gamma irradiation of southern US cultivars. Inheritance of these mutants was studied through progeny tests in M_5 to M_7 generations and through fertility characterization of pollen and spikelets at Stuttgart, Arkansas (Zhu and Rutger, 1999). For mutants Orion 1783 and Kaybonnet 1789 progenies from fertile panicles in segregating rows were all fertile; and progenies of near-sterile panicles segregated 1 fertile: 1 sterile, indicating that male sterility in each mutant was controlled by a single dominant gene. The other male sterile, Cypress 1819, provided a satisfactory fit to single recessive gene model, i.e., 3 fertile: 1 sterile in segregating generations. The two dominant male steriles showed partial pollen abortion; the recessive mutant, Cypress 1819, showed complete pollen abortion. Seed set on steriles under open pollination ranged from a low of 8.1% for Cypress 1819 to a high of 32.9% for Kaybonnet 1789. Bagged seed set of the three mutants which could be identified in the field at flowering, Orion 1783, Kaybonnet 1789, and Cypress 1819, averaged 3.5, 0.3, and 0%, respectively.

References

Oard, J.H., J. Hu, and J.N. Rutger, 1991. Genetic analysis of male sterility in rice mutants with environmentally influenced levels of fertility. Euphytica. **55**: 179-186

Rutger, J.N. and G.W. Schaeffer, 1994. An environmentally sensitive genetic male sterile mutant from anther culture of rice. In: Temperate Rice - Achievements and Potential. Humphreys, E., E.A. Murray, W.S. Clampett and L.G. Lewin (Eds.) Conf. Organizing Committee, NSW Agriculture, Griffith, Australia. pp.171-174

Shi, M., 1985. The discovery and study of the photosensitive recessive male-sterile rice (*Oryza sativa* L. subsp. *japonica*). Sci.Agri.Sin. **2**: 44-48

Yuan, S.-L., Z.-G. Zhang, H.-L Zen, K.-L Lu, J.-H. Lian, and B.-X. Wang, 1993. Two photoperiodic reactions in photoperiod-sensitive genetic male sterile rice. Crop Sci. **33**: 651-660

Zhu, X. and J.N. Rutger, 1999. Inheritance of induced dominant and recessive genetic male-sterile mutants in rice (*Oryza sativa* L.). SABRAO J. **31**: 17-22.

Aluminum tolerance in triticale, wheat, and rye*

B.Y. Kim[1], A.C. Baier[2], D.J. Somers[3], and J.P. Gustafson[4]

[1] *Department of Agronomy, Taegu University, Jinryang-myun, Kyungson-gun, Kyngbuk, Korea*

[2] *Centro Nacional de Pesquisa de Trigo, Empresa Brasileria de Pesquisa Agropecuária, (CNPT-EMBRAPA), Passo Fundo RS99001-970, Brazil*

[3] *Agriculture and Agri-Food Canada Saskatoon Research Centre, Saskatoon, Saskatchewan, S7N-0X2 Canada*

[4] *U.S. Department of Agriculture-ARS, Plant Genetics Research Unit, and Plant Science Unit, University of Missouri, Columbia, Missouri 65211, USA*

Key words: acid soil, wheat, rye, triticale, aluminium tolerance

Summary

Acid soils containing high levels of aluminum (Al) are known to severely limit plant growth on over 1.6 billion hectares worldwide. In the United States, a gradual decline in the pH of many soils both in the Great Plains as well as the Southeast, has caused many soils to become high in levels of free Al. This worldwide condition encouraged the analysis of wheat (*Triticum aestivum* L. em Thell.), triticale (X *Triticosecale* Wittmack), and rye (*Secale cereale* L.) germplasm from one of the major acid soil regions of the world (Brazil) in order to evaluate and compare the genetic potential of Al genes for cereal improvement. The objectives were to compare Al-tolerance levels in wheats, triticales, and ryes by measuring root elongation responses in Al-containing hydroponic nutrient solutions. Root elongation was impaired for all species grown in 1mg/L concentrations of Al. Rye had the longest root regrowth and Al-sensitive wheats had the shortest root regrowth. The triticales containing a 2D(2R) substitution developed in the mid-1970s had the poorest root regrowth of all triticale types. The newly developed advanced triticale lines (AABBRR) yet to be released for commercial production showed the highest degree of Al tolerance of all the triticale types and approach or exceed the levels observed in rye. This indicated that progress is being made in improving triticale's Al-tolerance in Brazil. Of all the old and new wheat varieties showing the highest degree of Al-tolerance, none of them were better than 'BH 1146' a variety that is at least 50 years old. This indicated that over the past 50 years, although Brazilian wheat breeders have made yield improvements in wheat production, they have not improved Al-tolerance. Rye showed a higher degree of Al-tolerance than the other cereals, when tested in 1 mg/L of al, but as expected some variation was noted.

* *This paper reports the results of research only. Mention of a proprietary product does not constitute an endorsement or a recommendation for its use by the USDA - ARS, and the University of Missouri. Contribution from the U. S. Department of Agriculture, Agricultural Research Service, Plant Genetics Research Unit, and the University of Missouri Agricultural Experiment Station; Journal Series No. 12,704.*

M. Maluszynski and K.J. Kasha (eds.),
Mutations, In Vitro and Molecular Techniques for Environmentally Sustainable Crop Improvement, 101–111.

Introduction

With the worlds population currently over 6 billion and currently growing at rate of about 1.9 to 3.9% per year (depending on the country involved), it is clear that existing agricultural production need to accelerate in order to keep up with population growth and an ever increasing demand for food. The United Nations has indicated that agricultural production will have to increase approximately 66% by 2040. Therefore, not only is a tremendous increase in food production on existing soils needed, but also new land under cultivation will be needed in order to meet the increase in food demand. The United Nations Food and Agricultural Organization has indicated that the last major land masses that can potentially be converted into agricultural production are the acid soils of the world (for example: those of the Cerrado in Brazil, and the Llanos of Colombia and Venezuela).

Acid soils are known to severely limit plant growth on about 1.6 billion hectares worldwide, which represents roughly 49% of all arable land (Granados *et al.*, 1993). A reduction in the use of liming has resulted in a further lowering of soil pH and increased aluminum (Al) toxicity, thus limiting production in many traditional wheat producing areas and the expansion of wheat production into nontraditional areas (Carver *et al.*, 1988; Briggs and Taylor, 1994). The development of new cultivars with increased Al-tolerance under both favorable and unfavorable environments is regarded as a fundamental and economic solution to helping increase world food production. Increased Al-tolerance should also make liming and fertilization more cost effective (Foy *et al.*, 1965; Foy and da Silva, 1991; Ruiz-Torrez *et al.*, 1992; Granados *et al.*, 1993; Briggs and Taylor, 1994). The value of using Al-tolerant cultivars is clearly evident from comparing near isogenic lines of wheat that differed in Al-tolerance; 31% more spikes, 66% more biomass, and 68% higher grain yield were obtained when Al-tolerant lines were grown on acid soil (Carver *et al.*, 1993).

Species and genotypes within species are known to differ widely in their tolerance to Al (Foy *et al.*, 1974; Little, 1988; Aniol and gustafson, 1990; Rajaram, *et al.*, 1991). Acid-soil and/or Al-tolerance has been identified in wheat (*Triticum aestivum* L. em Thell.) (Foy *et al.*, 1965; Mesdag and Slootmaker, 1969; Taylor and Foy, 1985; Carver *et al.*, 1988; Foy and da Silva, 1991; Nyachiro and Briggs, 1994; Aniol, 1995; Luo and Dvorak, 1996), and triticale (X *Triticosecale* Wittmack) and rye (*Secale cereale* L.) (Aniol *et al.*, 1980; Camargo and Felicio, 1984; Camargo *et al.*, 1991; Aniol, 1996) genotypes from various origins. Aniol *et al.* (1980) concluded that Al tolerance in rye inbred lines ranged between the Al-tolerant and Al-sensitive wheat cultivars. However, Camargo and Felicio (1984) reported that many rye cultivars tolerated significantly higher levels of Al in nutrient solution, than the Al-tolerant wheat and triticale cultivars.

Triticale germplasm containing a complete genome of rye chromosomes (AABBRR) has been shown to possess better adaptation and yield potential in marginal environments (Rajaram *et al.*, 1993). Many so-called 'complete' triticales showed Al-tolerance levels closer to wheat than rye (Gustafson and Ross, 1990). This is an indication that the expression of the genes in rye controlling Al tolerance appears to be somewhat suppressed when in a wheat background. It is implied that the improvement of Al tolerance in triticale may be achieved by either increasing the expression of rye Al tolerance gene(s), or by selecting increased Al tolerance in rye followed by combining the highly Al-tolerant ryes with Al-tolerant wheats.

The objectives of the current study were to evaluate the levels of Al tolerance in different triticale and wheat genotypes from Brazil and to compare the results to that of known rye genotypes using a simple and rapid screening technique utilizing Al-containing hydroponic solutions. The results of this evaluation will be related to progress being made in breeding Al-tolerant Brazilian wheats and triticales.

Materials and methods

Germplasm

The germplasm used in this study included two rye, 19 triticale, and nine wheat genotypes all from Brazil (Dr. A.C. Baier, EMBRAPA, Passo Fundo, Brazil) as well as one rye cultivar from Poland and one rye landrace variety from Ecuador courtesy of the USDA-Sears collection (Table 1). The Brazilian triticale genotypes were comprised of three groups: (1) hexaploid triticales cultivars not containing a complete rye genome, but containing a 2D(2R) substitution that was developed prior to 1980 in Brazil; (2) hexaploid triticale cultivars containing a complete rye genome (AABBRR) with no substitutions and were developed between 1980 and 1985 for cultivation in Brazil; and (3) advanced hexaploid triticale lines containing a complete rye genome (AABBRR) currently being evaluated in field trials for release in Brazil. The hexaploid wheat cultivars were comprised of two groups containing of both old and modern cultivars grouped according to their Al-tolerance: (1) sensitive; (2) tolerant based on replicated field trials conducted on acid soils in Brazil.

Aluminum stress and analysis

Seeds of each genotype were placed on moist filter paper in Petri dishes, held at 2±1°C for 12 h and were then germinated for 24 h at room temperature. Six seedlings of each genotype with similar root lengths (3 to 10 mm) and comparable endosperm sizes were selected in order to eliminate any variation in the result due to endosperm or root length differences at the start of the study. The selected seedlings were placed on plastic mesh floating on 2 L of an aerated, low ionic strength hydroponic medium [400 μM $CaCl_2$; 650 μM KNO_3; 250 μM $MgCl_2$; 10 μM $(NH_4)_2SO_4$; 40 μM NH_4NO_3 (pH 4.0)] containing Al (modified from Moore *et al.*, 1976; Aniol, 1991). The Al was added as $AlCl_3$ to the acidified hydroponic medium at concentrations of 0.5, 1.0, 2.0, and 4.0 mg Al/L (mg/L= 37 μM of Al). A 'zero Al' control was also included for each genotype. The solution pH was adjusted to 4.0±0.02 with HCl prior to, and with KOH, after the Al was added. No precipitation was observed in any of the solutions. The hydroponic tanks and seedlings were placed in a controlled environment (26°C±1°C 16 h day/8 h night) with a photon flux density of 1000 mmol $m^{-2}s^{-1}$. Seedling growth continued for 4 days with the solutions being changed each day in order to minimize potential changes in the pH and Al concentration.

The two longest roots from 6 seedlings for each of two replications were measured and averaged for each genotype treatment and for all genotypes within a germplasm type. The root tolerance index (RTI) of germplasm type and genotype was calculated by dividing the average root length in each Al concentration by the average root length in the solution with no Al (Taylor and Foy, 1985). Genotypes, as sub-plots, were randomized within each of two replications. Analysis of variance using a split-plot design was performed with germplasm type as the whole-plot factor and genotypes as the sub-plot factor.

It was previously determined that 1.0 mg/L Al produced the highest correlation between root growth and established acid-soil-tolerance levels (Baier *et al.*, 1996). Therefore, average root length and RTI values of the germplasm types and genotypes in the solutions containing 0.5 mg and 1.0 mg/L Al were compared by Fisher's LSD.

Table 1. Germplasm types, genotypes, pedigrees and the classification in to which each genotype was included

Germplasm type genotype (culivar)	Pedigree
Al-sensitive wheat cultivars	
Anahuac 75	II 12300//LR 64/7C/3/NOR 67
IAPAR 30	ALD Sib//CNT 7/PF 70354/3/ PAT 24//BB//KAL
Al- tolerant wheat cultivars	
Trigo BR 43	PF 833007/Jacui
Trigo BR 35	IAC 5*2/3/CNT7*3/LD//IAC5/HAD
EMBRAPA 15	CNT 10/BR 35//PF 75172/Tifton
Trigo BR 23	CC//ALD/3/IAS54-20/COP//CNT 8
CNT 1	PF 11.1000.62/BH 1146
Trigo BR 37	Mazoe/F 13279//Pelado Marau
IAC 5-Maringa	Frontana//Kenya 58/Ponta Grossa 1
BH 1146	Ponta Grossa 1//Fronteira/Mentana
Subtituted (2D/2R) triticale cultivars	
Triticale BR 1	M2A/CML (Panda)
Triticale BR 2	FS 3972-48M-0N-0Y-0F
CEP 18	TOB/8156//CC/3/Inia/4/SPY/5/M2A (Teddy)
IAPAR 13-Araucaria	M2A/CML//FN
Complete triticale cultivars	
Triticale BR 4	BGL/CIN//MUS
EMBRAPA 17	BGL/3/MTZTCL/Trigo//BGL/4/Nutria (Tatu)
EMBRAPA 18	Tapir/Yogui//2*MUS
CEP 22	BGL/CIN//IRA/BGL
CEP 23	BGL/3/MTZTCL/Trigo//BGL/4/Nutria (Tatu)
CEP 25	B6712-171-11Y-4Y-0M-0A
IAPAR 23-Arapoti	CIN/CNO//BGL/3/Merino (Hare)
IAPAR 38-Araruna	JLO/Panther
OCEPAR 3	CIN/CNO//BGL/3/Merino (Hare)
Complete triticale advanced lines	
PFT 8922	MUS/Lynx//Yogui/3/MUS
PFT 104	Tatu*2/China 7
PFT 107	Hare 263/Civet
PFT 222	LT-1/Rhino
TCEP 878	Hare 263/Civet
PR 884 (IAPAR54, OCEPAR4)	B6811-270-27Y-3Y-0M (Uron)
Rye genotypes	
Blanco	(Brazilian, selected for Al 5 cycles at U of M)
Centeio BR 1	(Cultivar from Brazil)
Dankowskie Zlote	(Cultivar from Poland)
Ecuador	(Landrace from Ecuador)

Results

The 19 triticale, nine wheat, and four rye cultivars were utilized to evaluate the Al tolerance levels, under hydroponic conditions, of triticales relative to wheat and rye, and to compare the results to previously conducted field trials. For this purpose, the differences for root elongation and RTI among Al concentrations, germplasm types (GT) for triticale, wheat, and rye, GTxAl interaction, GT (1 mg/L Al), and cultivars (CT) were measured (Table 2). The mean square differences in both root length and RTI for GT, GTxAl, and GT in 1.0 mg/L Al were all highly significant (p=0.001).

Table 2. Analysis of variance of root length and RTI values for wheat, triticale and rye genotypes following 4 days growth in hydroponic solutions with different Al concentrations

Sourse of variation	df	Mean squares Root length	df	Mean squares RTI
Replication (Rep)	1	103.9	1	72.9
A1	4	18020.8**	3	2653.3**
Rep x A1 (error A)	4	30.6	3	28.6
Germplasm type (GT)	5	1753.6**	5	2508.7**
Germplasm type x A1	20	304.8**	15	142.2**
GT (1 mg/L A1)	5	729.4**	5	823.9**
Cultivars (CT)	27	799.0**	27	934.0**
Cultivars x A1. GT	108	196.2**	81	132.8**
Cultivars (1 mg/L A1)	27	223.5**	27	934.6**
Error	160	23.7	128	37.2

** Significant at P=0.01

Root elongation of all genotypes was reduced in the lowest concentration of Al (0.5 mg/L), but the degree of reduction in root length varied among the GT and the cultivars within the GT (Fig. 1A; Table 2). The root growth in the control solution was not figured in any statistical analysis as it is obvious from Table 2 that any amount of A1 suppresses root growth of all species to some degree. In the control solution containing no Al, the Al-tolerant wheats had the longest roots, and the complete triticales and the Al-sensitive wheats had the second longest roots. The substituted triticales had the shortest roots of all the germplasm types. Rye showed an intermediate root length in the control solution as compared to the wheats and triticales. However, considerable variation was observed within.

Al-sensitive wheats had the shortest roots in the 1mg/L Al solution. The Al-sensitive wheats were completely separated from all other wheats and/or triticales in Al-containing solutions (Fig. 1A and 1B). The average root lengths of the Al-sensitive wheats, the Al-tolerant wheats, the substituted triticales, the complete triticales, the advanced triticale lines, and the ryes were 21.6 mm, 42.2 mm, 22.6 mm, 35.6 mm, 42.9 mm, and 42.7 mm in 1.0 mg/L Al, respectively (Table 2). The Al-tolerant wheats, the advanced triticale lines, and the ryes showed the highest degree of Al tolerance by having the longest root growth in the 1.0 mg/L Al solution, while the group of Al-sensitive wheats and the group of substituted triticales showed the lowest degree of tolerance to the 1.0 mg/L Al solution when compared to the other wheat and triticale groups.

In the 1.0 mg/L Al solution, average RTIs of the Al-sensitive wheats, the Al-tolerant wheats, the substituted triticales, the complete triticales, the advanced triticale lines, and the ryes were 37.3 %, 48.4 %, 47.6 %, 49.2 %, 62.4 %, and 66.6 %, respectively (Table 2). The advanced triticale lines and the ryes were not significantly different in their degree of Al tolerance at 1.0 mg/L, while the Al-sensitive wheats showed the lowest degree of Al tolerance. The Al-tolerant wheats, and the substituted and complete triticales were not significantly different in terms of Al tolerance in 1.0 mg/L Al.

The recently developed advanced triticale lines had the longest roots and the best RTI as compared with the two other triticale types (Fig. 1A; Table 2). The RTIs of the advanced triticale lines did not differ significantly from that of rye in both 0.5 and 1.0 mg/L Al.

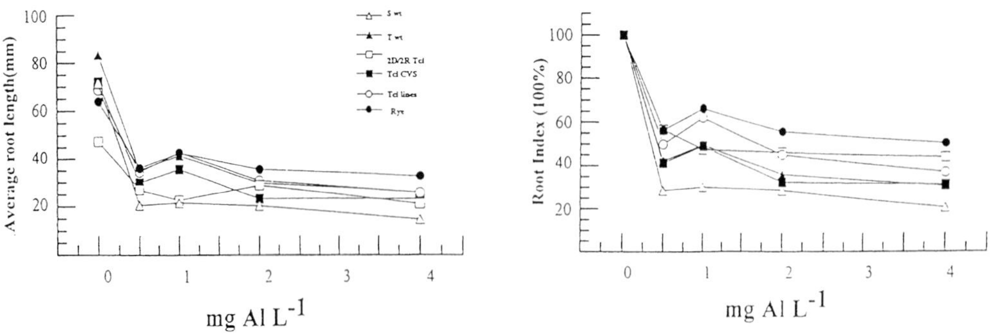

Figure 1. A. Root elongation of wheat, triticale, and rye germplasm types in various concentrations (mg/L) of Al-containing hydroponic solutions. Six seedlings of each genotype were grown for 4 d in hydroponic solutions. The average root length of each germplasm type was derived by averaging the two longest root measurements of all seedlings from each genotype within each germplasm type over two replications. Al-sensitive wheat cultivars, Al-tolerant wheat cultivars, substituted triticale cultivars, complete triticales, advanced triticale lines, and rye cultivars. B. Root tolerance index (RTI) of wheat, triticale, and rye germplasm types in various concentrations (mg/L) of Al-containing hydroponic solutions. Six seedlings of each genotype were grown for 4 d in hydroponic solutions. The RTI of each germplasm type was derived by averaging the two longest root measurements of all seedlings from each genotype within each germplasm type and two replications from the control solutions. Then dividing that by the two longest root measurements of all seedlings from each genotype within each germplasm type over two replications in various Al-containing hydroponic solutions. Al-sensitive wheat cultivars, Al-tolerant wheat cultivars, substituted triticale cultivars, complete triticales, advanced triticale lines, and rye cultivars.

Since Al-tolerance based on RTI expresses root growth relative to non-Al-containing solutions, as compared to total root growth following germination, the RTI values more accurately indicate a plant's ability to tolerate Al and continue root growth. The RTI values for the ryes, known to have a high degree of Al tolerance, and Al-sensitive wheats showed that they were clearly separated from all other germplasm types (Table 2). The substituted triticales had a slightly lower RTI than, but the difference was not significant, did the complete triticales, and both were significantly lower than the newly-developed advanced triticale lines Fig. 1B; Table 2).

The degree of variation in Al tolerance among all genotypes in wheat, triticale, and rye was large and resulted in three distinct significantly different group (Table 3). The complete triticale line 'IAPAR 38-Araruna' had the largest RTI (100.0 %) of all 33 genotypes studied at

Table 3. Average root lengths and RTI values of germplasm types and genotypes grown 4 days in 0.5 mg and 1 mg/L Al including control

Germplasm type genotype	Root length (mm)			RTI (%)	
	control	0.5 mg	1 mg	0.5 mg	1 mg
Sensitive wheat cvs.:	71.6	20.5A	21.6A*	37.4A	37.3A
Anahuac 75	29.0	15.0a	14.3a*	51.7	49.3
IAPAR 30	114.3	26.0b	29.0b	22.8	25.4
Tolerant wheat cvs.:	81.0	36.2	36.8	44.6	45.4
TrigoBR 43	80.3	37.8cd	37.0bc	47.0	46.1
TrigoBR 35	84.0	39.8cd	39.0c	47.3	46.4
TrigoBR 23	78.8	31.0bc	34.5bc	39.4	43.8
Highly-tolerant wheat cvs.:	90.8	34.5	46.2	38.4	50.6
BH 1146	114.8	43.0cd	62.0de	37.5	54.0
CNT 1	84.5	33.8bc	37.8bc	39.9	44.7
IAC5-Maringa	93.0	30.3bc	49.3c	32.5	53.0
Trigo BR 37	70.8	31.0bc	35.8bc	43.8	50.5
Subtituted triticale cvs.:	47.5	27.0B	22.6A	57.3C	47.6B
BR 1	31.8	19.8ab	14.0a	62.2	44.0
BR 2	48.5	27.0bc	25.0b	55.7	51.5
CEF 18	58.0	34.0bc	24.3b	58.6	41.9
IAPAR 13-Araucaria	51.8	27.3bc	27.3b	52.7	52.7
Complete triticale cvs	72.3	29.9B	35.6B	47.1B	49.2B
BR 4	55.8	32.3bc	28.8b	57.8	51.6
EMBRAPA 17	53.0	36.5c	31.5bc	68.9	59.4
EMBRAPA 18	107.0	25.2b	40.3bc	23.8	37.7
CEP 22	37.0	32.3bc	23.0b	87.2	62.2
CEP 23	77.0	25.8b	36.0bc	33.4	46.8
CEP 25	100.5	32.3bc	33.3bc	32.1	33.1
IAPAR 23-Arapoti	90.3	31.3bc	37.0bc	34.6	41.0
IAPAR 38-Araruna	40.3	19.8ab	40.3c	49.1	100.0
OCEPAR 3	90.0	33.3bc	50.3d	36.9	55.9
Advanced triticale lines:	68.8	34.4C	42.9C	55.8C	62.4C
PFT 8922	43.8	41.5c	40.9c	94.9	93.4
PFT 104	45.0	22.5ab	21.0ab	50.0	46.7
PFT 107	54.0	40.0cd	37.0bc	74.1	68.5
PFT 222	79.5	36.3c	39.3c	45.6	49.4
TCEP 878	102.5	33.0bc	63.3e	32.2	61.8
PR 884	87.8	33.3bc	56.3de	37.9	64.1
Rye cvs.:	64.1	36.1C	42.7C	57.3C	50.6C
Blanco	88.8	46.3d	55.0cde	52.1	61.9
Centeio BR 1	62.5	37.3cd	48.0cd	59.6	48.0
Dankowskie Zlote	66.0	36.3cd	44.8cd	54.9	67.9
Ecuador	93.3	24.5ab	23.0ab	62.4	24.7

*Averages of germplasm types followed by the same uppercase letter or averages of genotypes in all germplasm types followed by the same lowercase, do not differ by Fishers Protected LSD (P=0.05).

the 1.0 mg/L of Al. While wheat 'BH 1146' with an RTI of 54.0 and rye 'Dankowskie Zlote', originally from Poland, with an RTI of 67.9 % had the longest RTIs of the wheat and rye groups, respectively, in 1.0 mg/L level of Al. The Al-tolerance classes also showed variation within germplasm types for root length and RTI (Table 3).

Discussion

The germplasm evaluated represents a broad spectrum of old and modern cultivars present in triticale, wheat, and rye germplasm and cultivars currently being grown for production in Brazil. Both the triticales and wheats showed significant variation in response to Al in the hydroponic solutions (Table 2). The Al-sensitive wheats were clearly separated from other wheats, triticales, and ryes in response to Al.

The wheats were separated into the two groups, sensitive and tolerant, based on their known agronomic performance under field conditions in Brazil. The data indicated that 0.5 mg/L Al was not sufficient a concentration of Al to differentiate the Al-tolerance levels in wheat cultivars. However, an Al concentration of 1.0 mg/L was sufficient to separate the wheat groups giving the same rankings as the results observed in the field evaluations. The data showed that BH 1146 is still the most tolerant wheat in Brazil. This occurs in spite of the fact that BH 1146 has been a commercial cultivar in Brazil for over 50 years. This indicated that even though Brazilian wheat breeders have made significant progress in increasing wheat yields in the past 50 years, they have not germplasm to inter-cross into their programs that has improved levels on A1 tolerance. This suggests that there are probably no more genes available for increasing Al tolerance in the Brazilian wheat breeding programs. Since BH 1146 is also the most Al-tolerant spring wheat, so far screened, in the world (A. Aniol, pers. comm.) there might not be any additional un-utilized Al genes in the world's spring wheat germplasm for further wheat improvement in order to increase production on more acid soils. This has major implications for expanding world wheat acreage into more marginal acid soils (for example, those of the Cerrado in Brazil, and the Llanos of Colombia and Venezuela) in order to increase production.

The triticales were placed into three groups depending on age of the cultivar and chromosome composition in order to compare the cultivar/genotypes from different periods of time in the Brazilian breeding programs. On the whole, average root lengths of the three triticale groups, when stressed with Al, were intermediate between the sensitive wheats and highly-tolerant ryes. However, there were significant differences between the triticale groups. The average RTI of the advanced triticale lines was not significantly different from that observed in rye when stressed in an Al solution of 1.0 mg/L. The complete triticales produced in the mid-1980s had intermediate RTIs between the advance triticale lines and the substituted triticale cultivars released for commercial production before 1980. There was no significant difference between the RTIs of the substituted and complete triticale groups when tested in 1 mg/L Al. The advanced triticale lines had significantly larger RTIs than the other two triticale groups when stressed in an Al solution of 1.0 mg/L. This indicated that Brazilian triticale breeders have not only been improving yield and other agronomic traits between the mid-1970s to the 1990s, but they have also been gradually improving the level of Al tolerance in triticale beyond that observed in the most Al-tolerant wheat cultivar. The results indicate that the improved Al tolerance levels in triticale are coming from genes present in the rye genome of hexaploid triticale and not the genomes of wheat.

This is consistent with observations by Rajaram *et al.* (1993) who considered that triticale germplasm with a full complement of rye chromosomes showed better adaptation and yield potential in marginal environments in replicated field trials. The substituted triticales bred in the mid-1970s did tolerate a level of Al in solution comparable to the Al- tolerant wheats as was previously noted by Aniol *et al.* (1980). However, the substituted triticales were relatively poor performers on very acid soils, presumably due to a lack of additional rye Al-tolerance genes (Camargo and Felicio, 1984).

The root growth observed when rye germplasm was grown in Al-containing solutions is indicative of its known ability to grow on acid soils. The fast growing root system of rye is a particular trait of the species that might not necessarily relate to better Al tolerance. The differences in root elongation and RTI within the four ryes indicated that variation for Al tolerance exists within the small rye group analyzed and that further Al-tolerance improvement in rye as a species might be realized by intercrossing, screening, and selecting rye in hydroponic solutions containing Al.

Unfortunately, the expression of rye Al-tolerance genes in triticale is thought to be partially suppression by the interaction with the wheat genetic background. Evidence from rye addition and substitution lines into Al-sensitive wheats 'Anza' and 'Chinese Spring' demonstrated that genes on certain wheat chromosomes affected rye gene expression for Al tolerance (Aniol and Gustafson, 1984). Individual wheat chromosome arms were also shown to have a marked effect on rye Al-tolerance gene expression (Gustafson and Ross, 1990). Taken together, the data suggest that the Al-tolerance in wheat-rye hybrids is influenced by the genetic composition of both the wheat and rye parents (Gustafson and Ross, 1990). The significantly improved Al-tolerance levels observed in the advanced triticale lines of the present study indicated that the expression limitations on rye Al-tolerance gene expression can be progressively improved by breeding, whereas in wheat alone, there appears to be no additional genes for further improvement.

It would be of value to investigate further the observation that longer roots and higher RTI values of complete triticales were related to different genetic mechanisms in triticale (Rajaram *et al.*, 1991). If so, it may be possible to make significant improvements in the adaptation of triticale to acid soils not currently under cultivation. If triticales can be improved, then it should also be possible to transfer this rye genetic system for improved Al-tolerance into wheat backgrounds via chromosome substitutions and translocations, thus making an improvement in Al tolerance beyond that observed in the wheats BH 1146 and 'Atlas 66'.

New cultivars with improved Al tolerance will make wheat production on acid soils possible and more cost effective, thus providing an alternative to the current possibilities for improving production where acid soils exist. The current screening procedure to identify genotypes with improved acid-soil performance is rapid, cheap, and very useful for wheat and triticale breeding programs. Triticale breeders in Brazil appear to be selecting for additional Al tolerance from the rye parent, which suggests that by utilizing rye, Al-tolerance levels can be improved along with improvement of other agronomic traits. Since liming is becoming more expensive, Al-tolerant triticale and wheat cultivars will be preferred by farmers in the future as a stable means of increasing cereal production in many areas of the world where marginal soils low in pH and high in Al exist.

References

Aniol, A., 1991. Genetics of acid tolerant plant. In: Plant-Soil Interactions at Low pH. Wright, R.J., Baligar, V.C. and R.P. Murrmann (Eds.). Kluwer Acad.Pub., Dordrecht. pp.1007-1017

Aniol, A., 1995. Physiological aspects of aluminum tolerance associated with the long arm of chromosome 2D of the wheat (*Triticum aestivum* L.) genome. Theor.Appl.Genet. **91**: 510-516

Aniol, A., 1996. Aluminium uptake by roots of rye seedlings of differing tolerance to aluminium toxicity. Euphytica. **92**: 155-162

Aniol, A. and J.P. Gustafson, 1984. Chromosomal location of genes controlling aluminum tolerance in wheat, rye and triticale. Can.J.Genet.Cytol. **26**: 701-705

Aniol, A. and J.P. Gustafson, 1990. Genetics of tolerance in agronomic plants. In: Heavy Metal Tolerance in Plants: Evolutionary Aspects. Shaw, A.J. (Ed.) CRC Press, Boca Raton, Fl. pp.255-267

Aniol, A., R.D. Hill, and E.N. Larter, 1980. Aluminum tolerance of spring rye inbred lines. Crop Sci. **20**: 205-208

Baier, A.C., D.J. Somers, and J.P. Gustafson, 1996. Aluminum tolerance in triticale, wheat, and rye. In: Triticale: Today and Tomorrow. Guedes Pinto, H., N.L. Darvey and V.P. Carnide (Eds.) Kluwer Academic Publishers, Dordrecht. pp.437-444

Camargo, C.E.O. and J.C. Felicio, 1984. Tolerance of wheat, triticale and rye cultivars to different levels of aluminum in nutrient solution. Bragantia. **43**: 9-16 (English summary)

Camargo, C.E.O., J.C. Felicio, and A.W.P. Ferreira Filho, 1991. Triticale: aluminum tolerance in nutrient solution. Bragantia. **50**: 323-330 (English summary)

Carver, B.F., W.P. Inskeep, N.P. Wilson, and R.L Westerman, 1988. Seedling tolerance to aluminum toxicity in hard red winter wheat germplasm. Crop Sci. **28**: 463-467

Carver, B.F., W.E. Whitmore, E.L. Smith, and L. Bona, 1993. Registration of four aluminum-tolerant winter wheat germplasms and two near-isolines. Crop Sci. **33**: 1113-1114

Foy, C.D., W.H. Armiger, L.W. Briggle, and D.A. Reid, 1965. Differential aluminum tolerance of wheat and barley varieties in acid soils. Agron.J. **57**: 413-417

Foy, C.D. and A.R. da Silva, 1991. Tolerance of wheat germplasm to acid subsoil. J.Plant Nutr. **14**: 1277-1295

Foy, C.D., H.N. Lafever, J.W. Schwartz, and A.L. Fleming, 1974. Aluminum tolerance of wheat cultivars related to region of origin. Agron.J. **66**: 751-758

Granados, G., S. Pandey, and H. Ceballos, 1993. Response to selection for tolerance to acid soils in a tropical maize population. Crop Sci. **33**: 936-940

Gustafson, J.P. and K. Ross, 1990. Control of alien gene expression for aluminum tolerance in wheat. Genome. **3**: 9-12

Little, R., 1988. Plant soil interactions at low pH problem solving - the genetic approach. Commun.Soil Sci. **19**: 1239-1257

Luo, M.-C. and J. Dvorak, 1996. Molecular mapping of an aluminum tolerance locus on chromosome 4D of Chinese Spring wheat. Euphytica. **91**: 31-35

Mesdag, J. and L.A.J. Slootmaker, 1969. Classifying wheat varieties for tolerance to high soil acidity. Euphytica. **18**: 36-42

Moore, D.P., W.E. Kronstad, and R.J. Metzger, 1976. Screening wheat for aluminum tolerance. In: Plant Adaptation to Mineral Stress in Problem Soil. Wright, M.J. and S.A. Ferrari (Eds.) Cornell Univ. Press, Ithaca. pp.287-295

Nyachiro, J.M. and K.G. Briggs, 1994. Seedling root response of some Kenyan bread wheat cultivars grown in acid nutrient culture solution containing aluminum. Plant and Soil. **158**: 141-144

Rajaram, S., M.M. Kohli, and J. Lopez-Cesati, 1991. Breeding for tolerance to aluminum toxicity. In: Plant Soil Interactions at Low pH. Wright, R.J., Baligar, V.C. and R.P. Murrmann (Eds.). Kluwer Acad.Pub., Dordrecht. pp.1019-1028

Rajaram, S., G. Varughese, O. Abdalla, W.H. Pfeiffer, and M. van Ginkel, 1993. Accomplishments and challenges in wheat and triticale breeding at CIMMYT. Plant Breeding Abstracts. **63**(2): 131-139

Ruiz-Torrez, N.A., B.F. Carver, and R.L Westerman, 1992. Agronomic performance in acid soil of wheat lines selected for hematoxiline staining pattern. Crop Sci. **32**: 104-107

Snowden, K.C., K.D. Richards, and R.C. Gardner, 1995. Aluminum-induced genes. Induction by toxic metals, low calcium, and wounding and pattern of expression in root tips. Plant Physiol. **107**: 341-348

Taylor, G.J. and C.D. Foy, 1985. Mechanism of aluminum tolerance in *Triticum aestivum* L. (wheat). II. Differential pH induced by spring cultivars in nutrient solution. Aust.J.Bot. **72**: 702-706.

Evaluation of 'Anahuac' wheat mutant lines for aluminum tolerance

A. Tulmann Neto[1], C.E. de Oliveira Camargo[2], J. Lopes de Castro and A.W.P. Ferreira Filho[2]
[1] *Centro de Energia Nuclear na Agricultura, Universidade de São Paulo (USP), Caixa Postal 96, 13400-970 Piracicaba, SP, Brazil*
[2] *Instituto Agronômico, Caixa Postal 28, 13001-970, Campinas, SP, Brazil*

Key words: aluminium tolerance, acid soil, limed soil, induced mutations, wheat, gamma rays

Summary

Seeds from the aluminum sensitive cultivar 'Anahuac' were treated by gamma-ray radiation. Twenty eight selected Al^{+3} tolerant mutants were compared to the original Anahuac and the tolerant wheat cultivars IAC-24 and IAC-60 from 1994 to 1996 in acid (Capão Bonito) and limed (Monte Alegre do Sul and Tatui) soil field trials. Grain yield and agronomic characteristics were analyzed. Twenty six mutant lines yielded higher than the sensitive Anahuac cultivar in the acid soils of Capão Bonito. Under limed soil conditions, the mutants had a similar yield to the original sensitive cultivar. The majority of the mutants were similar in yield to the tolerant cultivars IAC-24 and IAC-60 under both conditions. Some of the mutants showed altered agronomic characteristics but even in these cases this will not limit its utilization. The results indicated that tolerant lines with good characteristics may be obtained from a susceptible cultivar by mutation induction, thus allowing cropping in conditions where Al^{+3} is a limiting factor.

M. Maluszynski and K.J. Kasha (eds.),
Mutations, In Vitro and Molecular Techniques for Environmentally Sustainable Crop Improvement, 113–120.

Introduction

Breeding for aluminum tolerant cultivars is among the main objectives of wheat (*Triticum aestivum* L.) improvement in the State of São Paulo (Camargo, 1993) since Al^{+3} sensitivity is an important limiting factor for yield in the majority of acid soils (Foy *et al.*, 1973). Therefore, Mexican cultivars with good agronomic characteristics, but highly sensitive to Al^{+3} in the soil, have been recommended to São Paulo State farmers only for limed soils (Felicio *et al.*, 1976, Camargo *et al.*, 1988). For example, the most sensitive to aluminum toxicity cultivar 'Anahuac' when tested in nutrient solution, did not show growth in the primary roots after treatment in solutions with 2 mg/liter of Al^{+3} (Camargo *et al.*, 1987).

Deficiency in an agronomic characteristic, such as aluminum tolerance may be corrected in a cultivar which has a series of other good traits by using chemical or physical mutagens. In Brazil, this strategy has been used in various crops and interesting new lines or cultivars of dry beans, soybean and chrysanthemum (Tulmann Neto and Sabino, 1994; Tulmann Neto and Latado, 1996b; Tulmann Neto and Alves, 1997) were obtained. Mutation induction *in vivo* or *in vitro* has resulted in the release of hundreds of cultivars in different crops, with mutants for a series of characteristics, including tolerance to abiotical stresses such as salinity, cold, high temperatures, herbicides and drought tolerance (Maluszynski *et al.*, 1995). Wheat mutation induction has been used in Brazil with several objectives, and mutants for earliness, plant height, disease resistance and Al^{+3} toxicity tolerance have been obtained (Tulmann Neto et al., 1995; Camargo *et al.*, 1996; Tulmann Neto *et al.*, 1996[a]). In the case of the sensitive cultivar Anahuac, the objective was to obtain Al^{+3} tolerance mutants without alteration to the basic characteristics of the cultivar following the previously described strategy. As in other crops, once the mutant is obtained, it is necessary to test its performance in various locations and years prior to deciding whether it may substitute for the original cultivar or will need to be intercrossed in a breeding program. (Barriga *et al.*, 1996) working with 26 gamma rays induced mutants in wheat concluded through trials in various locations and years that only one of them could be released as a cultivar in several regions with different environmental conditions in Chile.

Mutants obtained from gamma-ray irradiated seeds of the sensitive wheat cultivar Anahuac previously screened in nutrient solution for Al^{+3} tolerance were further selected under field conditions. The present study was carried out to assess the mutant lines and the original Anahuac, IAC-24 and IAC-60 cultivars in two locations in the São Paulo State for several years under field conditions. The evaluated traits were grain yield and agronomic characteristics.

Material and methods

As a result of earlier mutagenic treatments (Camargo *et al.*, 1996) in the Anahuac cultivar, sensitive to aluminum toxicity, 83 tolerant mutant lines were recovered. These lines were studied in 1992 in a screening nursery and based on agronomic characteristics, 28 lines were selected. In 1993 these were grown in preliminary trials at the locations of Monte Alegre do Sul and Tatuí. These were trials for agronomic trait evaluations and used for multiplication of seeds.

Two experiments were prepared. The first one was composed of 11 mutant lines obtained from earlier mutagenic treatments of the Anahuac cultivar, 6 selected lines of Anahuac, the original Anahuac and the wheat cultivars IAC-24 and IAC-60. The second experiment was composed of 17 mutant lines from earlier mutagenic treatments of Anahuac, the original Anahuac and the wheat cultivars IAC-24 and IAC-60 - both showing tolerance to aluminum toxicity.

A randomized block design with 4 replications was utilized for the two experiments. Each experiment was consisted of 80 plots, each plot composed of six rows 3 m long, spaced 0.20 m apart. Plots were separated from each other by 0,60 m. and 80 seeds were used per meter of line. Two locations were utilized for experiment 1 in 1994 and 1995: Capão Bonito represented by acid aluminum soil and upland condition, and Monte Alegre do Sul, under sprinkler irrigation and limed soil. Two locations were also used for the experiment 2 in 1994 and 1995: Capão Bonito, and Tatuí under sprinkler irrigation and limed soil. At the Capão Bonito station two seeding date trials for experiments 1 and 2 were carried out in 1995 and repeated at Capão Bonito in 1996. Besides grain yield the genotypes were evaluated for plant height, earliness and lodging.

Results

The result means of the four trials of experiment 1, carried out at Capão Bonito, in 1994, 1995 (seeding date 2) and 1996 are shown in Table 1. All mutant lines were tolerant to Al^{+3} toxicity (acidity) present in the soil and were significantly more productive compared to the original Anahuac which was sensitive to Al toxicity (acidity). The cultivar IAC-60, tolerant to Al^{+3} toxicity was the most productive, not differing from the cultivar IAC-24 nor from the Anahuac mutant lines L-2, L-3, L-4, L-6, L-9, L-10 and L-11.

In the two trials of experiment 1 carried out at Monte Alegre do Sul, in 1994 and 1995 (Table 2), the Anahuac mutant lines L-1, L-3, L-4, L-6, L-9 and L-10 were the most productive but did not differ significantly in grain yield from the original Anahuac nor from the controls IAC-24 and IAC-60. The genotypes L-3 and L-6 had a good grain stability in both locations (acid and limed soils) and were not much taller than the original Anahuac.

The results of four trials of experiment 2, carried out at Capão Bonito (acid soil and upland condition), in 1994, and 1995 (two seeding dates) and 1996 are presented at Table 3. All mutant lines, with exception of L-16 and L-17, which had exhibited tolerance to Al toxicity in nutrient solutions were more productive than the original Anahuac. The cultivar IAC-60 was the most productive genotype in these experiments.

Table 4 shows the result means of the three traits of experiment 2, carried out at Tatuí, in 1994 and 1995. All mutant lines, the original Anahuac and the cultivars IAC-24 and IAC-60 did not differ relative to grain yield. The mutant lines L-3, L-5, L-6, L-7, L-9, L-10, L-13 and L-14 which were the most productive in acid soil when compared with the original Anahuac (Table 3) did not differ in plant height nor days to flowering, with exception of L-3 which was taller.

Discussion

Barriga *et al.* (1990) in a study on wheat in Chile, observed that among 26 induced mutants for various characteristics, some had advantages over the control when tested for greater phosphorus absorption. The mutants were excellent phosphorus absorbers and outperformed the five controls in his experiments in regions of Chile with acid soil pH and high phosphorus fixation. Based on this result they recommended mutation induction for this characteristic.

These authors emphasized the need to establish yield stability by analyzing production for several years and locations before a final recommendation is made (Barriga *et al.*, 1996).

Table 1. Mean results of four trails of experiment 1, carried out at Capão Bonito, in 1994, 1995 (two seedings dates) and 1996

Selected lines	Cycle Emergence - flowering (days)	Emergence - - maturity (days)	Lodging (%)	Plant height (cm)	Grain yield (kg/ha)
Anahuac	58	126	0	41	123
Anahuac	57	126	0	44	151
Anahuac (original)	58	126	2	44	104
Anahuac	57	126	2	41	107
Anahuac	58	126	2	41	162
Anahuac	58	126	2	40	131
Anahuac	57	126	2	47	192
L-1	53	120	25	76	1064
L-2	56	121	12	67	1170
IAC-60	61	128	5	81	1662
L-3	55	120	22	78	1318
L-4	57	122	17	78	1318
L-5	55	120	15	70	1090
L-6	54	120	22	75	1409
L-7	50	118	15	70	1056
L-8	65	128	7	72	1078
L-9	51	119	18	84	1404
L-10	53	121	15	85	1426
L-11	51	118	18	83	1469
IAC-24	61	128	18	68	1208
F (genotypes)	5.79*	2.55*		12.62*	31.43*
L.S.D (Tukey) at 5%	8	12		26	554
CV%	5.53	1.59		12.52	23.26

In a recent review (Maluszynski *et al.*, 1995) observed that using mutation induction, hundreds of mutants of diverse crops, mainly cereals, have been released to farmers. Mutants obtained after mutagenic treatment are very often released directly to farmers after yield trials, but could also be included in breeding programs for new cultivars. In wheat production for bakeries, the same authors mention that only 35 out of 140 mutant cultivars have been obtained from indirect use in crosses. This means that in the majority of cases there are no large alterations induced by the mutagens in the original material. However, in several selected wheat mutants obtained in other countries, alterations have been found for other characteristics besides those that were the object of the original selection (Donini *et al.*, 1984). In many cases these characteristics are not undesirable and therefore do not limit direct use by farmers, but this indicates that care should be taken in assessing these mutants before deciding their use. In rye, a mutant was selected originally for low plant height, using a recessive monogenic mutation. Later it was found that this same gene was responsible for salinity tolerance both in the original mutant and in lines from crosses where it was included as parental. A positive pleiotropic effect of the gene was found to exist (Donini *et al.*, 1984; Forster *et al.*, 1995). But in other cases there may be a negative pleiotropic effect as happened with the TMD-1 dry bean mutant which is tolerant to golden mosaic virus but affects yield. Therefore, this mutant was not used directly but included in crossing programs (Tulmann Neto *et al.*, 1988) and it is a parent of the IAPAR-57 cultivar.

Table 2. Mean resutls of two trials of experiment 1 carried out at Monte Alegre do Sul, in 1994 and 1995

Selected lines	Cycle (days) Emergence - - flowering	Emergence - - maturity	Lodging (%)	Plant height (cm)	Grain yield (kg/ha)
Anahuac	73	116	15	86	2878
Anahuac	75	115	33	87	2827
Anahuac (original)	75	116	30	87	2826
Anahuac	75	115	55	88	2809
Anahuac	75	116	25	85	2927
Anahuac	75	115	48	87	2819
Anahuac	76	115	45	83	2694
L-1	69	114	68	112	2799
L-2	68	113	35	91	2712
IAC-60	77	125	18	96	3104
L-3	70	114	53	96	2938
L-4	70	115	60	98	2937
L-5	69	113	45	94	2676
L-6	69	113	53	94	2774
L-7	65	112	23	99	2389
L-8	75	120	3	86	1924
L-9	68	113	53	113	2972
L-10	68	114	55	114	3031
L-11	68	114	58	113	2649
IAC-24	76	120	8	86	2566
F (genotypes)	12.36*	5.96*		30.85*	2.42*
L.S.D (Tukey) at 5%	6	7		11	952
CV%	2.45	1.72		5.88	14.67

In the present work, tolerant mutants were found in the experiments. When compared with the original cultivar some of them were different for some agronomic characteristics (height, lodging, life cycle). But even in cases where the mutant was taller, which could be undesirable because of a greater lodging possibility, it was found that the lodging was low and did not differ in the majority of cases from the IAC-60 and IAC-24 cultivars which are recommended for the two types of soil used in the experiment. The experiments under acid soil conditions showed that 26 mutants yielded more than the original cultivar but under limed soil, the mutants were similar in yield to the sensitive cultivar. It is known that tolerance to aluminum toxicity in wheat is due to a pair of dominant genes (Camargo, 1993). Studies on the mutation genetics for tolerance are still in progress. But the results obtained in this study indicate that mutation for aluminum toxicity tolerance did not have negative pleitropic effects.

The results obtained on the assessed characteristics indicate that several of the mutants could be used for planting in acid or limed soils as a substitute for the original Anahuac cultivar. These results also indicate the validity of the technique for inducing variation for environment stress factors and, therefore, would be another case of induced mutants released directly for cultivation.

Table 3. Mean results of four trials of experiment 2 conducted at Capão Bonito in 1994, 1995 (two seedings dates) and 1996

Selected lines	Cycle Emergence - - flowering (days)	Cycle Emergence - - maturity (days)	Lodging (%)	Plant height (cm)	Grain yield (kg/ha)
L-1	59	134	10	68	1444
L-2	60	132	14	68	1424
L-3	54	125	13	91	1628
L-4	60	134	8	70	1344
L-5	54	129	8	73	1572
L-6	53	125	6	65	1549
L-7	52	126	5	67	1558
L-8	57	129	6	68	1425
L-9	58	129	5	69	1553
L-10	56	125	3	77	1444
L-11	52	129	4	71	1060
L-12	54	129	8	68	1021
L-13	56	125	13	73	1563
L-14	50	126	11	71	1547
L-15	50	125	10	67	1305
L-16	55	129	5	65	803
L-17	58	127	5	49	134
Anahuac (original)	58	127	4	51	286
IAC-60	62	134	5	81	1937
IAC-24	60	134	9	72	1503
F (genotypes)	9.70*	2.06*		13.94*	17.14*
L.S.D (Tukey) at 5%	6	12		12	566
CV%	5.40	2.57		8.33	19.83

Conclusions

- Twenty six mutant lines obtained by gamma-ray radiation from the Anahuac cultivar were higher yielding than the original Anahuac cultivar in acid soil under upland conditions. The majority of these lines had a yield similar to the IAC-60 and IAC-24 cultivars, which are tolerant to aluminum toxicity.
- The majority of the mutant lines and the controls Anahuac, IAC-60 and IAC-24 did not differ among themselves for grain yield in lime corrected soils under sprinkler irrigation.
- Some of the mutant lines had altered agronomic characteristics (plant height, cycle, lodging) when compared to the original Anahuac cultivar, but such modifications did

not diminish their grain yield. These mutants may be utilized as a substitute for the Anahuac cultivar.
- These results demonstrated the efficiency of mutation induction techniques for increasing the genetic variability for yield limiting factors such as aluminum toxicity.

Table 4. Mean resutls of two trials of experiment 2 carried out at Tatuí, in 1994 and 1995

Genotypes	Lodging (%)	Plant heigth (cm)	Grain yield (kg/ha)
L-1	5	84	3254
L-2	7	83	3167
L-3	4	105	3344
L-4	24	85	3156
L-5	2	85	3764
L-6	4	80	3483
L-7	24	79	3649
L-8	14	81	3660
L-9	4	79	3573
L-10	0	86	3629
L-11	0	84	3788
L-12	21	86	3830
L-13	21	83	3456
L-14	37	81	3760
L-15	16	82	3431
L-16	2	85	3299
L-17	1	85	2896
Anahuac (original)	2	84	3125
IAC-60	10	94	3292
IAC-24	4	83	3323
F (genotypes)		6.92*	1.34
L.S.D (Tukey) at 5%		12	885
CV%		5.02	12.64

Acknowledgments

The authors thank the Cooperatives of the Paranapanema Valley, the FAPESP and the International Atomic Energy Agency, Vienna, Austria for supporting this study under the FAO/IAEA Research Contract 7641.

References

Barriga, P., R. Fuentes, P. Seemann, and N. Manquian, 1990. Variabilidad genetica de mutants de trigo en la absorcion y utilizacion de fosforo. Turrialba. **40**: 279-286

Barriga, P., R. Fuentes, P. Seemann, and N. Manquian, 1996. Yield stability of wheat mutant lines. In: Use of Mutation Techniques for Improvement of Crops in Latin America. IAEA-TECDOC-859. IAEA, Vienna. pp.141-148

Camargo, C.E.O., 1993. Trigo. In: Melhoramento de plantas no Instituto Agronomico de Campinas. Furlani, A.M.C. and G.P. Viegas (Eds.) Instituto Agronomico, Campinas. pp.433-488

Camargo, C.E.O., J.C. Felicio, A.W.P. Ferreira Filho, C. Barros de, A. Pettinelli Junior, R.R. Santos, R.A.D. Kanthack, and L.S. Rocha Junior, 1988. Melhoramento do trigo. XVII Comportamento de linhagens de origem mexicana no Estado de Sao Paulo. Bragantia. **47**(1): 25-41

Camargo, C.E.O., J.C. Felicio, A.W.P. Ferreira Filho, A. Pettinelli Junior, A. Tulmann Neto, and A. Ando, 1996. Mutation breeding of wheat in S. Paulo State, Brazil. In: Use of Mutation Techniques for Improvement of Crops in Latin America. IAEA-TECDOC-859. IAEA, Vienna. pp.95-111

Camargo, C.E.O., J.C. Felicio, and L.S. Rocha Junior, 1987. Trigo: Tolerancia ao aluminio em solucao nutritiva. Bragantia. **46**(2): 183-190

Donini, B., T. Kawai, and A. Micke, 1984. Spectrum of mutant characters utilized in developing improved cultivars. In: Selection in Mutation Breeding. IAEA, STI/PUB/665, Vienna. pp.7-31

Felicio, J.C., C.E.O. Camargo, and C. Barros de, 1976. Estudo comparativo de cultivares de trigo em latossolo roxo no Estado de Sao Paulo em 1974. Bragantia. **35**: 147-154

Forster, B.P., H. Packnyat, C.G. Simpson, and L.L. Handley, 1995. Genetic control of salt tolerance in barley. In: Induced Mutations and Molecular Techniques for Crop Improvement. IAEA, Vienna. pp.347-343

Foy, J.C., A.L. Fleming, and J.W. Schwartz, 1973. Opposite aluminum and manganese tolerances of two wheat varieties. Agr.J. **65**: 123-126

Maluszynski, M., B.S. Ahloowalia, and B. Sigurbjörnsson, 1995. Application of *in vivo* and *in vitro* mutation techniques for crop improvement. Euphytica. **85**(1-3): 303-315

Tulmann Neto, A. and M.P. Alves, 1997. Induction of mutations for earliness in the soybean cultivar Parana. Brazil.J.Genet. **20**(1): 45-50

Tulmann Neto, A., C.E.O. Camargo, M.P. Alves, J.L. Castro, and P.B. Gallo, 1995. Inducao de mutacao visando a reducao de altura de planta e resistencia as doencas, no cultivar de trigo (*Triticum aestivum* L.) IAC-17. Scientia Agricola. **52**(2): 287-293

Tulmann Neto, A., C.E.O. Camargo, M.P. Alves, A. Pettinelli Junior, and A.W.P. Ferreira Filho, 1996. Reducao na altura de planta e resistancia as doencas no cultivar de trigo (*Triticum aestivum* L.) IAC-19 atraves de inducao de mutacoes. Revista Brasileira de Genetica. **19**(2): 275-281

Tulmann Neto, A. and R.R. Latado, 1996. Cristiane and Ingrid - first chrysanthemum cultivars obtained by induced mutations in Brazil. MBNL. **42**: 18

Tulmann Neto, A., J.O.M. Menten, A. Ando, J. Alberini, and T.C. Peixoto, 1988. Use of induced mutations for disease resistance and other agronomic charcteristics in beans (*Phaseolus vulgaris* L.) and soybean (*Glycine max* (L.) Merrill). In: Improvement of grain legume production using induced mutations. IAEA, STI/PUB/766, Vienna. pp.167-188

Tulmann Neto, A. and J.C. Sabino, 1994. Inducao e uso de mutante de habito de crescimento determinado e precoce em feijoeiro (*Phaseolus vulgaris* L.). Revista Brasileira de Genetica. **17**(4): 425-430.

Barley mutants with increased tolerance to aluminium toxicity

M. Nawrot[1], I. Szarejko[1] and M. Maluszynski[2]
[1] *Department of Genetics, Silesian University, Jagiellonska 28, 40-132 Katowice, Poland*
[2] *Plant Breeding and Genetics Section, Joint FAO/IAEA Division, IAEA, Vienna, P.O.Box 200, 1400-Vienna, Austria*

Key words: abiotic stress, aluminium tolerance, genetic analysis, *Hordeum vulgare*, mutagenesis, screening methods

Summary

Acid soil and associated aluminium toxicity are considered as the number one abiotic factor limiting crop production. Over 2 billion hectares of acid soils exist world-wide, both in tropical and moderate climatic zones. In Poland acid soils represent up to 60% of arable land. At soil pH < 5.0 Al ions become soluble in water and toxic to plants. Genetic improvement of Al tolerance in crops is the only alternative to soil liming, a traditional but short term and expensive agricultural cure to raise soil pH. Of the various cereals, barley is the most sensitive to Al toxicity. The known sources of Al tolerance in barley are limited to old cultivars and landraces. While they represent multiple alleles of a single locus, there is no potential to improve Al tolerance through recombination of non-allelic additive genes. In the Department of Genetics, Silesian University we have employed induced mutations for rapid creation of variability for Al tolerance in barley. Thirteen mutants with increased levels of tolerance to Al toxicity have been selected in M_3 generation after mutagenic treatment of four barley varieties with N-methyl-N-nitroso urea (MNH) and sodium azide. Six further Al tolerant mutants were identified in the collection of semi-dwarf mutants of the Department. All selected mutants confirmed Al tolerance with the use of three different methods of screening, i.e., root re-growth, root tolerance index and hematoxylin staining. Fourteen mutants exhibited significant root re-growth after 48 hour incubation with 3 ppm Al^{+3} and two of them, namely RL819/2 and RL820/6 were tolerant even to 6 ppm Al^{+3}. Crosses of two selected mutants with their respective parent varieties indicated that Al tolerance in each mutant was controlled by a single recessive gene. Out of three methods tested, the root re-growth method facilitated by hematoxylin staining proved to be the most reliable technique for large scale testing. Double treatment with MNH or combined treatment with sodium azide and MNH and 6h inter-incubation germination between treatments were the most successful treatment combinations for induction of aluminium tolerance in barley.

M. Maluszynski and K.J. Kasha (eds.),
Mutations, In Vitro and Molecular Techniques for Environmentally Sustainable Crop Improvement, 121–137.

Introduction

Plant sensitivity to toxic effects of aluminium ions is one of the major factors limiting crop production in acid soils, which represent up to 40 % of the world's arable land (Haug, 1984). Over 2 billion hectares of acid soils exist world-wide, both in tropical and moderate climatic zones. It is estimated that 391 million hectares in Europe are highly acidic. In Poland, about 60 % of the cultivated land is classified as acid, with 24 % having a pH below 4.5 and 36 % with a pH between 4.5 and 5.5 (Aniol, 1989). Aluminium is the most abundant metal in earth's crust, and weathering of Al-containing minerals under acidic conditions at pH < 5.0 allows aluminium ions (Al^{+3}) to become mobilized, soluble in water and available to plants (Wright, 1989; Delhaize and Ryan, 1995). The content of soluble Al ions in soils ranges from 1-33 ppm, but seldom exceeds 4 ppm. Increased concentrations of aluminium (for many plant species as low as 1-2 ppm) inhibits root cell division and elongation, resulting in poor root development, reduction of water and nutrient uptake, drought susceptibility, and subsequently in a significant decrease in yield (Foy, 1983). These negative effects can be partly overcome by soil liming, a traditional agricultural practice used to raise the soil pH and causing the Al to become insoluble. Conventional liming methods treat only the topsoil, while the mobilized aluminium in the subsoil (>1m depth) remains unaffected. Thus, a serious aluminium problem remains for any deeply-rooted crops, among them are the cereals.

An alternative way to lessen the effects of acid soil on crop production is the cultivation of Al tolerant crops that sustain high yields on these marginal soils, making soil amendment unnecessary or utilizing them more effectively. Species and genotypes within species differ widely in their tolerance to aluminium as various plants use different strategies to avoid toxic effects of aluminium ions (Foy, 1988; Aniol and Gustafson, 1990). The production of Al tolerant varieties provides a long-term solution that can be adapted to most regions as an economically and ecologically sound practice.

Of the various cereal species, barley is the most sensitive to Al toxicity and even micromolar concentrations of Al inhibit root growth (Aniol, 1989). The known sources of Al tolerance in barley are limited to old cultivars and landraces. A detailed study performed on 37 barley genotypes of diverse origin indicated that only 1 locus with multiple alleles was responsible for Al tolerance in barley (Minella and Sorrells, 1992). This study was in agreement with earlier results of Reid (1971) who found that increased Al tolerance in barley was inherited as a single dominant gene. Thus, the improvement of Al tolerance in barley cannot be achieved through recombination of non-allelic additive genes. Minella and Sorrells (1992) concluded that the known genetic sources of Al tolerance had a limited value for barley breeding and a search for new genes/alleles was urgently needed. This search can be performed either through further screening of existing germplasm or through employment of mutation techniques. Since all known sources of Al tolerance are alleles of a single locus, induced mutations seem to be a choice for rapid creation of desired variability.

Mutation techniques have contributed significantly to world-wide plant improvement. According to the FAO/IAEA Mutant Varieties Database, more than 1840 mutant varieties involving 164 plant species have been officially released up to December 1997, and some of them have made an outstanding impact on the productivity of a particular crop (Maluszynski *et al.*, 1995a; Maluszynski *et al.*, 1999). In some countries, mutant varieties of economically important crops, e.g., barley, durum wheat and cotton, occupy the majority of cultivated areas. Semi-dwarfness and earliness are plant characters most often improved by mutagenic treatment, but there are also examples of successful mutagenic induction of tolerance to abiotic stresses, such as salt tolerance in barley (Forster et al., 1995) and rice (J. Deus, person.

commun.), or cold tolerance in rice (Maluszynski et al., 1995b). Mutation techniques have also been successfully used for the induction of Al tolerance in wheat. After gamma irradiation of the Al sensitive cultivar 'Anahuak', 28 Al tolerant mutant lines were obtained and 3 of them performed better than a recently released variety, both in acid and limed soil (Camargo *et al.*, 1995).

Screening for aluminium tolerance in acid soil in the field has many obvious limitations, among them a difficulty in dealing with large batches of soil with the same concentration of Al ions needed for large-scale selection. For this reason several protocols to screen for Al toxicity in nutrient media have been developed. While the primary effect of aluminium toxicity is the inhibition of root growth, most of these protocols are based on one of the three approaches: 1) the comparison of the root elongation in a nutrient solution with a toxic level of Al^{+3} to that in the same solution without aluminium (Baier *et al.*, 1995; Khatiwada *et al.*, 1996), 2) the ability of plants to re-start root growth after a brief period of exposure to the toxic concentration of Al^{+3} (Camargo, 1984; Aniol, 1991), and 3) the estimation of the damage caused by accumulation of Al^{+3} in the root tips based on their stainability with some histochemical indicators which form complexes with Al, such as hematoxylin, morin or Eriochrome cyianine R (Polle *et al.*, 1978; Minella and Sorrells, 1992; Luo and Dvorak, 1996; Larsen *et al.*, 1998). Additionally, the protocols proposed by authors for the same species often differed in: the content and concentration of macro- and micro-salts in the nutrient medium; the concentration of Al^{+3}; the duration of plant growth in a nutrient solution; the time of the exposure to Al stress; and the pH of the nutrient solution itself (pH 4.0 or pH 4.5). These differences make it difficult to compare the experimental results obtained with the use of one specific screening method with data published by other authors.

The objective of this study was to produce new sources of Al tolerance in barley utilizing induced mutations of commercial varieties cultivated in Poland and to identify a mode of inheritance of increased tolerance to Al toxicity in the selected barley mutants. Different methods of screening for Al tolerance in barley were examined in order to establish a protocol that allows one to analyse a large number of plants in a short time.

Material and Methods

Plant material

For selection of Al^{+3} tolerant forms, M_3 seedlings from available mutated barley populations were used. The mutated material was derived from chemical mutagenesis of four spring barley varieties ('Dema', 'Magda', 'Maresi', 'Rudzik'). In total, 67,870 M_3 seedlings, representing more than 6,787 individual M_2 plants were examined for aluminium tolerance (Table 1). This material was derived from different mutagenic combinations, including double treatment (2 x 3h) with different doses of N-methyl-N-nitrosourea (MNH), or combined treatment with sodium azide (NaN_3) and MNH, and 6h inter-incubation germination period between treatments (Maluszynska and Maluszynski, 1983; Szarejko and Maluszynski, 1999). In a separate study, 371 semi-dwarf (sd) mutants from the collection of the Department of Genetics, Silesian University were evaluated for Al tolerance. The sd mutants represented M_{15} and further generations after mutagenic treatment with MNH, NaN_3, gamma rays and fast neutrons of 12 barley varieties ('Aramir', 'Delisa', 'Diva', 'Georgia', 'HDM', 'Julia', 'Karat', 'Mg 4170', 'Plena', 'Roland', 'Salka', 'Trumph'). Additionally, two barley lines (CB 240 and CB 245) selected as Al^{+3} tolerant in Brazil and provided by the courtesy of Dr. C.E. Camargo from the Agronomy Institute, Campinas, Sao Paulo were included in the experiments.

Table 1. Mutated populations used for selection of Al^{+3} tolerant mutants in barley.

Variety	Treatment combination	No. of M_3 seedlings analysed	No. of corresponding M_2 plants	No. of corresponding M_1 plants
Dema	0.5 mM MNH - 6h iig - 0.5mM MNH	8510	851	100
	0.75 mM MNH - 6h iig -0.75 mM MNH	2810	281	50
	1.5 mM NaN_3 - 6h iig- 0.75 mM MNH	9180	918	100
Magda	0.5 mM MNH - 6h iig - 0.5 mM MNH	3070	307	50
	1.5 mM NaN_3 - 6h iig- 0.75 mM MNH	3620	362	201
Maresi	0.5 mM MNH	5850	585	100
	1.0 mM MNH	3790	379	100
	1.5 mM MNH	5600	560	62
	1.5 mM NaN_3 - 6h iig - 0.75 mM MNH	9180	918	100
Rudzik	0.5 mM MNH - 6h iig - 0.5 mM MNH	13260	1326	150
	1.5 mM NaN_3 - 6h iig- 0.75 mM MNH	3000	300	50
Total number of plants analysed		67870	6787	1063

Screening methods for Al tolerance

The initial screening for barley mutants tolerant to aluminium was performed with the use of the root re-growth method proposed by Camargo (1984), modified at our Department (Protocol 1) through the addition of hematoxylin staining as proposed by Polle et al. (1978). The technique estimates the ability of plant roots to re-grow in a nutrient solution without Al^{+3} after a period of growth in the solution with toxic levels of aluminium. The staining with hematoxylin facilitates selection of the tolerant forms. The Al^{+3} concentration of 0.75 ppm was chosen for the first selection to ensure that all potentially tolerant forms can be detected.

Ten M_3 seedlings per M_2 plant were used for the initial screening with 0.75 ppm Al^{+3}. Mutant lines showing significantly increased tolerance were evaluated again at 1 ppm Al^{+3} and checked for homozygosity in M_4 generation.

The comparative study with two others methods used for assessing aluminium tolerance was performed for selected mutants:

1. Root tolerance index (RTI), the relative root elongation in the presence and absence of Al in hydroponic nutrient solutions (Baier *et al.*, 1995), with modifications (Protocol 2).
2. Hematoxylin staining (Polle *et al.*, 1978; Minella and Sorrells, 1992) (Protocol 3). This method evaluates the level of Al^{+3} sensitivity/tolerance on the basis of hematoxylin stainability of roots after growth of seedlings in three concentrations

of Al^{+3}: 0.03 mM, 0.06 mM and 0.09 mM (equivalent to 0.81 ppm, 1.62 ppm, 2.43 ppm Al^{+3}).

M_4 mutants or sd mutant lines from the collection showing significant tolerance to 1 ppm Al^{+3} in all three applied methods were evaluated again with the use of higher concentrations of Al^{+3} (2-4 ppm). Additionally, tolerance to 5 and 6 ppm concentrations of Al^{+3} ions was evaluated for mutants indicating root re-growth at 4 ppm Al^{+3}. In all experiments 3 replications were used with 10 seedlings per replication. Protocols for three methods applied in the experiments and composition of nutrient media are presented below.

Protocol 1

Modified root re-growth method of screening for Al^{+3} tolerance (Camargo, 1984 and Polle *et al.*, 1978)

1. Seeds are pre-germinated in Petri dishes with wet filter paper (3-5 layers), first in a refrigerator for 48 hours and then at room temperature for the next 5-6 hours, until the coleorhiza appears.
2. Germinated seeds are plated into floating trays, covered with black plastic film and placed in containers with aerated nutrient solution “A” (Table 1). About 30 ml of solution per seedling is used. The containers are placed in a growth chamber at 22-24°C, 3000 lux, 16^h photoperiod.
3. After 48 hours, the floating trays are moved into Al^{+3} containing nutrient solution “B” (Table 2). It is 10x less concentrated than solution “A” and because of amelioration of aluminium toxicity by phosphorus, there is no KH_2PO_4 in this solution. The nutrient medium “A” used in the first part of experiment is stored for the further use.
4. After 48 hours of growth in the Al^{+3} containing nutrient solution "B", roots are rinsed with distilled water and floating trays are transferred back into nutrient solution “A”.
5. After the next 72 hours, roots are rinsed again with distilled water and stained with 0.2% hematoxylin solution. To prepare 1 L of hematoxylin solution, 2g of hematoxylin and 0.2 g $NaIO_3$ are dissolved in 1000 ml distilled water.
6. Results of staining: parts of the roots that accumulated Al^{+3} have a dark navy-blue colour. Plants tolerant to aluminium continue growth and below the stained part of the root, white re-growth of the root tip is visible.

Protocol 2

Modified root tolerance index (RTI) method of screening for Al^{+3} tolerance (Baier *et al.*, 1995)

1. Seeds are pre-germinated in Petri dishes with wet filter paper (3-5 layers), first in refrigerator for 24 hours and then in an incubator at 24°C for 12 hours.
2. Germinated seeds with seminal roots about 3 mm long are transferred into floating trays. Trays are placed in containers with aerated low ionic strength hydroponic medium, containing Al^{+3} (Table 4). One seedling requires ~ 50 ml of solution. The control without Al^{+3} is included in each experiment. Containers are put in a growth chamber at 22-24°C, 3000 lux, 16^h photoperiod.

3. During the whole experiment, the pH must be constant 4.0 ± 0.05. pH should be checked every day and adjusted, if necessary to 4.0 ±0.05 with HCl
4. After 4 days of plant growth, the length of the two longest roots from each seedling is measured.
5. Root tolerance index (RTI) is calculated as the ratio of average root length in each Al^{+3} concentration to the average root length of the same genotype grown in the solution without Al^{+3}.

Table 2. Nutrient medium composition for the root re-growth method. Solution "A" (pH 4.0± 0.05)

Components	Final concentration	Stock solution* 200 x (g/L)
Macro salts	mM	
$Ca(NO_3)_2 \times 4H_2O$	4.0	188.9
$MgSO_4 \times 7H_2O$	2.0	98.6
KNO_3	4.0	80.9
$(NH_4)_2SO_4$	0.435	11.6
KH_2PO_4	0.50	13.6
Micro salts	µM	
$MnSO_4 \times H_2O$	2.0	0.068
$CuSO_4 \times 5H_2O$	0.3	0.016
$ZnSO_4 \times 7H_2O$	0.8	0.046
NaCl	30.0	0.350
$Na_2MoO_4 \times 2H_2O$	0.1	0.0046
H_3BO_3	10.0	0.124
Iron source	µM	
$FeNa_2EDTA$	10.0	0.73

*use 5 ml of stock solution per 1 L medium

Protocol 3

Hematoxylin staining method of screening for Al^{+3} tolerance (Polle *et al.*, 1978 and Minella and Sorrells, 1992)

1. Seeds are pre-germinated in Petri dishes with wet filter paper (3-5 layers), first in refrigerator for 48 hours and then at room temperature for the next 5-6 hours, untill coleorhiza appears.
2. Germinated seeds are placed into floating trays. They are covered with black plastic film and placed in containers with aerated nutrient solution (Table 5). One seedling requires ~ 30 ml of solution. The container are placed in a growth chamber at 22-24°C, 3000 lux, 16^h photoperiod. After 24 hours of growth, pH is adjusted to 4.0±0.05 with HCl. Seedlings are grown in the nutrient solution without Al^{+3} for the next 30 hours.
3. After 30h of growth, Al^{+3} is added to the nutrient solution. Each genotype should be evaluated at 0.03 mM, 0.06 mM and 0.09 mM concentration of Al^{+3}. Seedlings are grown in Al^{+3} containing solutions for 17 hours.

4. After 17 hours of growth in Al^{+3} solution, the seedlings are rinsed and kept in distilled water for 60 min.
5. Roots are stained with hematoxylin staining solution for 15 min.
6. After staining, roots are washed briefly and returned to distilled water for 30 min.
7. Root tips are scored for the degree of hematoxylin staining at the 0.03, 0.06, 0.09 mM Al^{+3} (0.81, 1.62 and 2.43 ppm). Those parts of roots that accumulated Al^{+3} have a dark navy-blue colour. At each Al concentration approximately 5 mm of root tip of a genotype is scored as: completely stained (C), partially stained (P), or non staining (N). Additionally, the distance of the not stained part of the root can be measured under stereo-microscope. The scores of root stainability corresponding to the level of Al^{+3} tolerance, according to Minella and Sorrels (1992) are presented in Table 6.

Table 3. Nutrient medium composition for the root re-growth method. Solution "B" (pH 4.0± 0.05; Al^{+3} is added as $AlCl_3$ x 6 H_2O)

Components	Final concentration	Stock solution* 200 x (g/L)
Macro salts	mM	
$Ca(NO_3)_2$ x4H_2O	0.4	188.9
$MgSO_4$ x 7H_2O	0.2	98.6
KNO_3	0.4	80.9
$(NH_4)_2SO_4$	0.0435	11.6
Micro salts	µM	
$MnSO_4$ x H_2O	0.2	0.068
$CuSO_4$ x 5H_2O	0.03	0.016
$ZnSO_4$ x 7H_2O	0.08	0.046
NaCl	3.0	0.350
Na_2MoO_4 x 2H_2O	0.01	0.0046
H_3BO_3	1.0	0.124
Iron source	µM	
$FeNa_2EDTA$	1.0	0.73

*use 5 ml of stock solution per 1 L medium

Table 4. Nutrient solution for the root tolerance index method (pH 4.0 ± 0.05;.Al^{+3} is added as $AlCl_3$ x 6 H_2O)

Components	Final concentration (µM)	Stock solution* 1000x (g/L)
$CaCl_2$ x 2H_2O	400	44.4
KNO_3	650	65.7
$MgCl_2$	250	21.8
$(NH_4)_2SO_4$	10	1.32
NH_4NO_3	40	3.2

*use 1 ml of stock solution per 1 L medium

Genetic analysis of Al tolerant mutants

In order to identify the genetic mechanism controlling increased tolerance to Al toxicity, the selected barley mutants were crossed with the parent variety and F_1, F_2, F_3 generations were produced. F_3 seedlings derived from individual F_2 plants (30 seedlings/F_2 plant) were evaluated for Al tolerance together with the mutant, parent variety and F_1 generation. The segregation ratios of Al tolerant to sensitive plants in F_2 generation were calculated on the basis of F_3 screening. Using this approach it was possible to score each F_2 plant as homozygous or heterozygous for Al tolerance/sensitivity. The screening was performed with the root re-growth method, at 1 ppm Al^{+3}. Three replications were used with ten seedlings per replication for each evaluated plant.

Table 5. Nutrient solution for the hematoxylin staining method (pH 4.0±0.05; Al^{+3} is added as $AlCl_3 \times 6H_2O$)

Components	Final concentration (mM)	Stock solution* 100x (g/L)
$CaCl_2$	4.0 mM	44.40
KNO_3	6.5 mM	65.70
$MgCl_2$	2.5 mM	23.80
$(NH_4)_2SO_4$	0.1 mM	1.32
NH_4NO_3	0.4 mM	3.20

*use 10 ml of stock solution per 1 L medium

Table 6. Hematoxylin stainability of barley root tips related to Al tolerance (Minella and Sorrels, 1992)

Hematoxylin stainability scores*	Al tolerance level
CCC	Very sensitive (VS)
NCC, PCC	Moderately sensitive (MS)
NPC, PPC	Moderately tolerant (MT)
NNN, NNP, NPP, PPP	Tolerant (T)

*C – complete staining; P – partial staining; N – no staining

Results and discussion

In total, 41 M_3 putative mutant lines derived from 4 barley varieties showed increased tolerance to 0.75 ppm Al^{+3} in the initial screening with the root re-growth method. Thirteen of these mutants were confirmed to have increased tolerance to the higher level, 1 ppm of Al^{+3} ions, in both M_3 and M_4 progeny (Table 7). These mutant lines were selected in M_3 progeny of two modern Polish varieties: Dema (5 mutants) and Rudzik (8 mutants) after mutagenic treatment with chemical mutagens (MNH and sodium azide). The mutants were found in the M_3 progeny of 1032 of Dema and 1626 of Rudzik M_2 plants (Table 1), indicating a relatively high frequency of M_1 plants carrying mutation(s) responsible for increased Al^{+3} tolerance. On the other hand, none of the putative mutants from varieties Maresi and Magda, selected as tolerant to 0.75 ppm Al^{+3} in M_3 generation, were confirmed as tolerant to the higher level (1 ppm) of Al^{+3} ions, neither in the same nor the next generation (Table 7). This result shows high genomic specificity of barley varieties to mutagenic action of the applied chemical agents, in relation to this character. High mutagenic potential of N-methyl-N-nitrosourea and sodium azide has been proven many times by several authors (Warner *et al.*, 1977; Maluszynski *et al.*, 1988; Ullrich and Aydin, 1988; Kaushik and Khush, 1991; Jende-Strid, 1993; Satoh *et al.*, 1997). At the Department of Genetics, Silesian University, Katowice, both MNH and sodium azide have been routinely used for mutation induction in barley. Mutants exhibiting different changes in root system development were found in mutated barley generations with the similar high frequency of 3.3% (Maluszynski, 1999).

Table 7. Results of screening for Al^{+3} tolerant M_3 mutant lines with the use of modified root re-growth method - Protocol 1

Variety/Mutant	Root re-growth length (mm) at different Al^{+3} concentration			Variety/Mutant	Root re-growth length (mm) at different Al^{+3} concentration		
	0.75 ppm	1.0 ppm	1.0 ppm		0.75 ppm	1.0 ppm	1.0 ppm
	M_3	M_3	M_4		M_3	M_3	M_4
Dema	0.5±0.4	0.3±0.3	0.3±0.3	Rudzik	0.6±0.3	0.5±0.7	0.2±0.2
DM170/8	4.5±1.0*	0.9±0.2*	0.7±0.3*	RD65/6	5.2±3.7*	0.0±0.0	0.0±0.0
DM184/7	5.8±3.0*	0.3±0.4	0.3±0.6	RD65/7	5.8±2.5*	0.1±0.1	0.0±0.0
DM185/7	3.8±3.0*	0.0±0.0	0.0±0.0	RD65/8	4.9±3.5*	0.1±0.1	0.0±0.0
DM185/8	3.3±1.0*	0.0±0.0	0.0±0.0	RD66/2	6.8±3.5*	0.0±0.0	0.0±0.0
DM186/1	11.5±3.1*	3.0±2.8*	1.2±0.8*	RD85/6	5.8±1.0*	0.5±0.7	0.0±0.0
DM186/8	8.0±2.2*	2.0±1.4*	2.7±1.5*	RD114/1	6.0±2.9*	0.8±0.4	1.5±1.3*
DM189/6	9.9±2.3*	1.5±1.5	1.7±1.2*	RD119/5	4.5±1.0*	0.5±0.6	0.0±0.1
DM206/9	4.3±1.5*	0.0±0.0	0.0±0.0	RD128/1	6.5±2.4*	0.0±0.0	0.0±0.0
DM214/6	3.5±1.0*	0.0±0.0	0.0±0.0	RD138/1	6.2±2.5*	0.5±0.7	0.3±0.6
DM215/4	2.3±0.5*	0.0±0.0	0.0±0.0	RD139/2	6.8±4.3*	0.3±0.4	0.3±0.6
DM218/5	4.8±2.9*	0.0±0.0	0.0±0.0	RD145/8	7.8±3.8*	1.5±0.7*	2.7±2.1*
DM221/2	9.5±1.3*	1.7±1.8	1.3±0.6*	RD167/3	6.0±3.9*	2.0±1.4*	1.3±0.6*
Maresi	0.2±0.2	0.0±0.0	0.2±0.3	RD167/6	10.7±4.0*	1.0±0.0*	1.3±0.6*
MR211/5	3.0±1.4*	0.0±0.0	0.0±0.0	RD169/4	11.0±3.9*	3.5±2.1*	1.3±0.6*
MR211/7	3.5±1.9*	0.0±0.0	0.0±0.0	RD170/3	6.8±1.7*	1.0±1.4	0.3±0.6
MR213/13	3.1±2.3*	0.0±0.0	0.0±0.0	RD170/8	4.3±1.5*	0.0±0.0	0.0±0.0
MR218/1	3.5±1.0*	0.0±0.0	0.0±0.0	RD177/6	6.5±1.3*	0.0±0.0	0.0±0.0
MR220/1	4.8±0.5*	0.0±0.0	0.0±0.0	RD177/9	11.0±1.8*	2.0±1.4*	1.3±0.6*
MR250/13	6.0±2.9*	0.0±0.0	0.0±0.0	RD178/1	5.8±1.5*	0.3±0.4	0.3±0.6
Magda	0.4±0.4	0.1±0.1	0.1±0.1	RD180/6	9.5±3.7*	1.5±0.7*	1.7±1.1*
MD134/9	3.8±1.0*	0.0±0.0	0.0±0.0	RD180/12	6.8±3.3*	0.5±0.7	0.3±0.6
				RD190/6	8.3±2.5*	2.5±2.1*	2.3±2.3*

* - significantly better than parent variety at P=0.05

Two mutants from variety Dema (DM186/1 and DM186/8) originated from the same M_1 plant what suggests that they could be derived from the same mutagenic event. That was also the case of two mutants from variety Rudzik namely RD167/3 and RD167/6. All other mutants were derived from different M_1 plants.

Among semi-dwarf mutants from the collection of the Department of Genetics, 17 mutants expressed higher levels of tolerance to 0.75 ppm Al^{+3} than the parent variety and 8 of these forms were also tolerant to 1 ppm Al^{+3} (Table 8). The highest number of sd mutants with increased tolerance to Al^{+3} was found among mutants originated form the variety Roland. It should be noted, however, that Roland itself expressed increased tolerance to Al toxicity, as compared to other varieties analysed in this study. Four mutants derived from the variety Roland, namely RL806/9, RL807/8, RL808/5 and RL809/5 were obtained after gamma radiation and two mutants (RL819/2 and RL820/6) after chemical mutagenesis.

Table 8. Results of screening for Al^{+3} tolerant sd mutants by the modified root re-growth method - Protocol 1

Variety/Mutant	Root re-growth length (mm) at different Al^{+3} concentration	
	0.75 ppm	1.0 ppm
Aramir	0.8±0.4	0.1±0.1
043AR	4.3±1.0*	0.0±0.0
Diva	1.1±1.2	0.3±0.4
200DV	9.0±1.4*	0.0±0.0
203DV	6.5±2.1*	0.0±0.0
216DV	6.5±2.1*	0.0±0.0
225DV	6.0±1.4*	2.5±0.7*
237DV	4.0±1.4*	0.0±0.0
HDM	0.8±0.4	0.2±0.2
358HD	3.5±2.1*	0.0±0.0
Karat	0.7±0.5	0.0±0.0
678Q	3.5±2.1*	0.9±0.2*
709Q	3.0±1.4*	0.5±0.4
842Q	6.5±0.7*	0.8±0.4
Roland	2.3±0.4	1.3±0.4
RL806/9	4.4±0.6*	3.4±1.3*
RL807/8	9.0±1.4*	4.3±0.4*
RL808/5	9.3±1.0*	4.1±0.8*
RL809/5	4.6±0.6*	3.1±0.8*
RL819/2	11.5±2.1*	5.4±0.6*
RL820/3	4.5±0.9*	0.4±0.3
RL820/6	8.8±1.1*	6.0±0.5*

* - significantly better than parent variety at P=0.05

Tolerance to the higher concentration of Al^{+3} ions (2-6 ppm) was evaluated for the selected 19 mutants with the use of Protocols 1 and 2. All mutants derived from variety Dema and three mutants originated from variety Rudzik showed significant root re-growth up to the concentration of 3 ppm Al^{+3} (Table 9). Root tolerance index was significantly higher than in the parent variety even at a concentration of 4 ppm of Al^{+3} ions for all, except one, mutants of variety Rudzik and three mutants of variety Dema (Table 10). The highest level of Al tolerance was expressed by mutants derived from the variety Roland. All Roland mutants exhibited root re-growth at 4 ppm Al^{+3} and three of them were able to re-start root growth even at 6 ppm Al^{+3}, the highest concentration tested (Table 9).

Comparison of two protocols of screening for Al tolerance in barley, namely root re-growth and RTI methods, showed that the latter technique was less selective than the root re-growth method. For example, with the use of the RTI protocol, all 6 mutants from the variety Roland showed a significantly increased tolerance over the parent variety at 6 ppm Al, while only three of them (RL807/8, RL819/2 and RL820/6) were able to re-start their root growth at this concentration (Table 9 and 10). The RTI method, although shorter than the root re-growth procedure (4 and 7 days, respectively), requires much more labour in measurements and data evaluation. This could be a limitation in fast screening of tolerant individuals. The root re-growth method also allows for the immediate transfer to soil of recovered mutants.

In order to compare the level of Al tolerance of barley mutants selected in this study with literature references, the third method of evaluation was applied. This technique is based on hematoxylin staining ability of plant roots grown for 17 hours in solution with three different doses of Al, namely: 0.03 mM (0.81 ppm), 0.06 mM (1.62 ppm) and 0.09 mM (2.43 ppm). These are the doses recommended for evaluation of barley germplasm by Minella and Sorrells (1992). All mutants selected in the presented experiments can be scored as tolerant or moderately tolerant, according to the scale presented by the authors (Table 11). Among forms scored as tolerant, there were six mutants selected from variety Roland, five mutants derived from variety Rudzik and two mutants from variety Dema. Roots of these mutants exhibited no hematoxylin stainability at the 0.03 mM Al^{+3}, no or partial stainability at 0.06 mM Al^{+3}, and partial stainability at 0.09 mM Al^{+3}. These results correspond to the tolerance of known sources of aluminium tolerance in barley examined by Minella and Sorrels (1992). Mutants derived from varieties Rudzik and Roland and two barley lines, CB240 and CB245 selected as aluminium tolerant in Brazil, were additionally checked for hematoxylin stainability at 0.12 mM Al (3.24 ppm). Also at this higher concentration of Al ions, four mutants (two from Rudzik: RD180/6, RD190/6 and two from Roland: RL819/2 and RL820/6) were scored as partially stained. Two of them, RL819/2 and RL820/6 exhibited also the highest level of tolerance during evaluation with the two other methods used in this study, i.e., root re-growth and RTI (Table 9 and 10). At the same concentration 0.12 mM Al, both barley lines from Brazil were evaluated as completely hematoxylin stained, so they appear to be less tolerant than mutants RL819/2 and RL820/6. Out of three methods of selection for Al tolerance in barley, the root re-growth method facilitated by the hematoxylin staining can be recommended as the simplest and most reliable technique for large scale screening. A similar approach but with the use of different nutrient media and staining was used by Aniol (1991) for investigation of Al tolerance in wheat, rye and triticale.

Even though the methodology and Al^{+3} doses applied in this experiment were the same as described by Minella and Sorrells (1992), it is difficult to relate directly results of our study to the evaluation of barley germplasm presented by those authors. The most tolerant forms reported by them were scored as partially stained at 0.09 mM Al^{+3} but all exhibited complete staining at 0.12 mM. It is likely that mutants RL819/2 and RL820/6 which were scored as partially stained at 0.12 mM Al^{+3} in our study, exhibit higher levels of aluminium tolerance than tolerant forms described up to now in barley, but this conclusion remains to be checked in a separate experiment.

Genetic analysis of Al tolerance was performed for two selected mutants: RL819/2 from the variety Roland and RD177/9 from variety Rudzik, using the root re-growth method at 1 ppm Al^{+3} in the nutrient solution. The F_1 generation of crosses between mutants and their respective parent varieties showed that aluminium tolerance in both mutants was inherited as a

Table 9. Tolerance of selected barley mutants to different concentrations of Al^{+3} ions according to the modified root re-growth method – Protocol 1

Genotype	Root re-growth lenght (mm) at different Al^{+3} concentration					
	1 ppm	2 ppm	3 ppm	4 ppm	5 ppm	6 ppm
Dema	0.26±0.05	0.09±0.02	0.04±0.02	0.0 ± 0.0	-	
DM170/8	0.91±0.17*	0.23±0.07*	0.13±0.06*	0.03±0.02*	-	
DM186/1	0.88±0.05*	0.17±0.02*	0.19±0.02*	0.02±0.01	-	-
DM186/8	1.08±0.08*	0.19±0.06*	0.13±0.03*	0.02±0.02	-	-
DM189/6	1.24±0.14*	0.20±0.03*	0.15±0.05*	0.02±0.01	-	-
DM221/2	1.08±0.18*	0.16±0.05*	0.15±0.04*	0.01±0.01	-	-
Rudzik	0.19±0.08	0.02±0.01	0.03±0.06	0.0 ± 0.0	-	-
RD114/1	1.01±0.14*	0.18±0.02	0.17±0.08	0.0 ± 0.0	-	-
RD145/8	1.01±0.22*	0.77±0.10*	0.59±0.13*	0.0 ± 0.0	-	-
RD167/3	1.24±0.14*	0.76±0.15*	0.75±0.15*	0.0 ± 0.0	-	-
RD167/6	0.84±0.20*	0.21±0.10*	0.10±0.09	0.0 ± 0.0	-	-
RD169/4	1.02±0.16*	0.51±0.20*	0.43±0.12*	0.0 ± 0.0	-	-
RD177/9	0.81±0.07*	0.36±0.03*	0.17±0.05	0.0 ± 0.0	-	-
RD180/6	1.03±0.15*	0.48±0.04*	0.11±0.02	0.0 ± 0.0	-	-
RD190/6	1.04±0.1*	0.53±0.15*	0.17±0.06	0.0 ± 0.0	-	-
Roland	1.26±0.49	0.09±0.10	0.02±0.02	0.01±0.01	0.0 ±0.01	0.0 ± 0.0
RL806/9	4.37±0.72*	1.71±0.70*	0.48±0.18*	0.34±0.26*	0.09±0.04	0.03±0.02
RL807/8	3.72±0.79*	1.45±0.61*	0.36±0.09*	0.29±0.10*	0.11±0.02	0.10±0.04*
RL808/5	3.60±1.59*	1.54±0.58*	0.30±0.11*	0.21±0.13*	0.19±0.09*	0.0 ± 0.0
RL809/5	3.26±1.19*	1.52±0.51*	0.65±0.15*	0.39±0.25*	0.22±0.17*	0.0 ± 0.0
RL819/2	5.53±1.74*	2.79±1.24*	0.67±0.29*	0.40±0.13*	0.43±0.07*	0.09±0.04*
RL820/6	7.66±2.32*	4.96±0.70*	1.87±0.29*	1.68±0.24*	1.44±0.17*	0.38±0.13*
CB 240	6.50±1.21*	0.04±0.04	0.05±0.05	0.09±0.07	0.01±0.01	0.0± 0.0
CB 245	1.33±0.47	0.69±0.13	0.15±0.05	0.14±0.15	0.01±0.01	0.0 ± 0.0

* - significantly better than parent variety at P=0.05; lines CB 240 and CB 245 were compared to the variety Roland

Table 10. Tolerance of selected barley mutants to different concentrations of Al^{+3} ions according to the modified root tolerance index (RTI) method - Protocol 2

Genotype	Root Tolerance Index at different Al^{+3} concentration					
	1 ppm	2 ppm	3 ppm	4 ppm	5 ppm	6 ppm
Dema	0.33±0.1	0.28±0.03	0.31±0.05	0.28±0.04	-	-
DM170/8	0.39±0.04	0.30±0.04	0.34±0.05	0.30±0.04	-	-
DM186/1	0.44±0.01*	0.40±0.02*	0.44±0.06*	0.34±0.07	-	-
DM186/8	0.42±0.03	0.33±0.07	0.46±0.04*	0.41±0.04*	-	-
DM189/6	0.53±0.02*	0.44±0.02*	0.51±0.03*	0.48±0.04*	-	-
DM221/2	0.42±0.04	0.35±0.04	0.45±0.06*	0.40±0.05*	-	-
Rudzik	0.33±0.04	0.23±0.03	0.26±0.02	0.17±0.001	-	-
RD114/1	0.41±0.07	0.25±0.03	0.37±0.02*	0.22±0.001	-	-
RD145/8	0.47±0.05*	0.31±0.06	0.39±0.06*	0.32±0.03*	-	-
RD167/3	0.67±0.03*	0.50±0.06*	0.46±0.05*	0.35±0.06*	-	-
RD167/6	0.50±0.03*	0.40±0.04*	0.37±0.02*	0.38±0.02*	-	-
RD169/4	054±0.09*	0.50±0.10*	0.47±0.13*	0.28±0.05*	-	-
RD177/9	0.56±0.06*	0.41±0.03*	0.37±0.07*	0.28±0.02*	-	-
RD180/6	0.57±0.13*	0.42±0.02*	0.34±0.01	0.33±0.03*	-	-
RD190/6	0.64±0.01*	0.56±0.09*	0.51±0.01*	0.46±0.02*	-	-
Roland	0.29±0.05	0.25±0.04	0.22±0.05	0.27±0.02	0.16±0.02	0.10±0.02
RL806/9	0.64±0.08*	0.47±0.08*	0.57±0.15*	0.49±0.07*	0.52±0.08*	0.63±0.04*
RL807/8	0.60±0.09*	0.51±0.04*	0.37±0.06*	0.37±0.05*	0.40±0.05*	0.39±0.06*
RL808/5	0.54±0.07*	0.47±0.07*	0.36±0.04*	0.36±0.07*	0.31±0.04*	0.32±0.09*
RL809/5	0.59±0.09*	0.55±0.09*	0.46±0.04*	0.42±0.07*	0.31±0.04*	0.29±0.01*
RL819/2	0.68±0.08*	0.54±0.04*	0.49±0.13*	0.51±0.10	0.50±0.07*	0.52±0.00*
RL820/6	0.71±0.04*	0.65±0.08*	0.55±0.03*	0.52±0.03*	0.47±0.07*	0.50±0.04*
CB 240	0.52±0.03*	0.49±0.13*	0.36±0.05	0.22±0.02	0.30±0.02*	0.34±0.01*
CB 245	0.31±0.02	0.25±0.01	0.31±0.05*	0.27±0.04	0.30±0.05*	0.40±0.04*

* - significantly better than parent variety at P=0.05; lines CB 240 and CB 245 were compared to the variety Roland

recessive trait (Table 12). The segregation for Al tolerant and Al sensitive plants in the F_2 generation were done on the basis of F_3 progeny tests that allowed easy identification of all F_2 genotypes. The data presented in Table 13 indicate that increased level of Al tolerance in both mutants was controlled by a single recessive gene. Therefore, these new genetic sources of Al tolerance appear to be different from those previously described as dominant at the analysed concentration of Al^{+3} by Minella and Sorrells (1992).

Al tolerant barley mutants in our study were obtained both in chemical and physical mutagenesis (Table 14). The most successful mutagenic combinations include double treatment with the same dose (0.5 or 0.7 mM MNH) and 6 hours of inter-incubation germination between treatments (Maluszynska and Maluszynski, 1983). Two mutants expressing the highest level of Al tolerance, namely RL819/2 and RL820/6 were obtained after double applications of 0.7 mM MNH. Also combined treatments with sodium azide and MNH gave good results as 7 mutants from different varieties resulted from this combination.

Table 11. Tolerance of selected barley mutants to different concentrations of Al^{+3} ions according to the hematoxylin staining method - Protocol 3

Genotype	Hematoxylin stainability scores at different Al^{+3} concentration				Length of non stained part of root (mm) at different Al^{+3} concentration				Result*
	0.03 mM	0.06 mM	0.09 mM	0.12 mM	0.03 mM	0.06 mM	0.09 mM	0.12 mM	
Dema	N	C	C		0.93±0.21	0.32±0.29	0.05±0.08		MS
DM170/8	N	P	C		1.66±0.50	0.38±0.15	0.10±0.09		MT
DM186/1	N	P	P		1.10±0.50	0.47±0.06	0.44±0.13		T
DM186/8	N	P	C		1.25±0.38	0.80±0.24	0.22±0.26		MT
DM189/6	N	N	P		1.26±0.43	0.86±0.10	0.43±0.10		T
DM221/2	N	P	C		1.72±0.60	0.83±0.33	0.13±0.10		MT
Rudzik	P	P	C	C	3.17±0.74	0.28±0.15	0.10±0.10	0.0±0.0	MS
RD114/1	P	P	C	C	2.75±0.96	0.94±0.56	0.27±0.39	0.17±0.41	MS
RD145/8	N	P	C	C	3.25±0.42	2.2±0.45	0.79±0.36	0.36±0.47	MT
RD167/3	P	P	C	C	3.10±0.82	1.88±0.35	0.71±0.49	0.55±0.46	MT
RD167/6	N	P	P	C	4.36±0.75	0.64±0.27	0.86±0.24	0.13±0.22	T
RD169/4	N	N	P	C	2.80±0.84	1.56±0.86	1.16±0.29	0.76±0.31	T
RD177/9	N	N	P	C	5.60±0.49	1.21±0.57	1.00±0.50	0.57±0.35	T
RD180/6	N	N	P	P	4.67±1.03	1.67±0.41	1.12±0.36	0.76±0.25	T
RD190/6	N	N	P	P	5.00±0.89	1.50±0.64	0.94±0.50	1.13±0.38	T
Roland	N	P	C	C	1.63±0.48	0.79±0.27	0.71±0.42	0.09±0.12	MT
RL806/9	N	N	P	C	4.00±0.93	4.25±0.46	0.86±0.24	0.40±0.35	T
RL807/8	N	P	P	C	5.70±1.16	3.00±0.70	2.63±0.91	0.19±0.19	T
RL808/5	N	P	P	C	4.50±0.76	3.50±0.72	1.50±0.53	0.10±0.14	T
RL809/5	N	N	P	C	4.20±1.08	3.30±1.30	1.88±1.11	0.29±0.37	T
RL819/2	N	N	P	P	5.50±0.54	5.50±0.55	1.83±0.40	0.88±0.25	T
RL820/6	N	N	P	P	6.80±1.50	5.20±0.79	1.57±0.53	0.78±0.30	T
CB240	N	P	P	C	2.60±0.99	1.80±0.94	1.14±0.38	0.96±0.46	T
CB245	N	N	P	C	2.20±0.83	2.45±1.46	1.13±0.35	1.20±0.58	T

*MS - moderately sensitive; MT - moderately tolerant; T - tolerant

Al^{+3} doses: 0.03 mM = 0.81mg; 0.06 mM = 1.62 mg; 0.09 mM = 2.43 mg; 0,12 mM = 3,24 mg

Table 12. Al tolerance of barley mutants RL819/2, RD177/9, their parent varieties Roland and Rudzik and F_1 progeny of the crosses between parents and mutants, according to the root re-growth method

Genotype	Root re-growth length (mm)	Comments
Roland	0.87±0.36	
RL819/2	3.60 ± 0.22	
F_1 Roland x RL819/2	0.95 ± 0.29	as Roland
Rudzik	0.21 ± 0.19	
RD177/9	1.90 ± 0.27	
F_1 Rudzik x RD177/9	0.15 ± 0.08	as Rudzik

Table 13. Segregation for Al^{+3} sensitive and tolerant plants in F_2 generation of the crosses Roland x RL819/2 and Rudzik x RD177/9 (on the basis of root re-growth of F_3 seedlings)

Cross-combination	F_2 genotype	No. of F_2 plants	Root re-growth length in F_3 (mm)	$\chi^2_{1:2:1}$
Roland x RL819/2	Homozygous sensitive	22	0.85 ± 0.38	0.613
	Heterozygous	47		
	Homozygous tolerant	19	3.80 ± 0.45	
Rudzik x RD177/9	Homozygous sensitive	78	0.18 ± 0.11	3.882
	Heterozygous	120		
	Homozygous tolerant	74	2.30 ± 0.25	

$P = 0.05\ \chi^2_{1:2:1} = 5.991$

Table 14. Successful mutagenic treatment combinations in developing Al tolerant mutants in barley

Mutant	Treatment combination
DM170/8, DM186/1, DM186/8, DM189/6, RD114/1, RD145/8	0.5 mM MNH - 6h iig - 0.5 MNH
DM221/2, RD167/3, RD167/6, RD169/4, RD177/9, RD180/6, RD190/6	1.5 mM NaN_3 - 6h iig - 0.75 mM MNH
RL819/2, RL820/6	0.7 mM MNH - 6h iig - 0.7 mM MNH
RL806/9, RL807/8, RL808/5, RL809/5	120 Gy gamma rays

References

Aniol, A., 1989. Principles of breeding cereals tolerant to low soil pH value. Biuletyn IHAR. **171/172**: 215-221 (in Polish with English summary)

Aniol, A., 1991. Genetics of acid tolerant plant. In: Plant-Soil Interactions at Low pH. Kluwer Academic Publishers, Dordrecht. pp.1007-1017

Aniol, A. and J.P. Gustafson, 1990. Genetics of tolerance in agronomic plants. In: Heavy Metal Tolerance in Plants: Evolutionary Aspects. Shaw, A.J. (Ed.) CRC Press, Boca Raton, Fl. pp.255-267

Baier, A.C., D.J. Somers, and J.P. Gustafson, 1995. Aluminium tolerance in wheat correlating hydroponic evaluations with field and soil performances. Plant Breed. **114**: 291-296

Camargo, C.E.O., 1984. Tolerancia de cultivares de trigo a diferentes niveis de aluminio em solucao nutritiva e no solo. Bragantia. **40**: 21-31

Camargo, C.E.O., A.W.P. Ferreira Filho, and A. Tulmann Neto, 1995. Genetic diversity in wheat and breeding for tolerance to acid soils. In: Induced Mutations and Molecular Techniques for Crop Improvement. IAEA, Vienna. pp.321-333

Delhaize, E. and P.R. Ryan, 1995. Aluminium toxicity and tolerance in plants. Plant Physiol. **107**: 315-321

Forster, B.P., H. Packnyat, C.G. Simpson, and L.L. Handley, 1995. Genetic control of salt tolerance in barley. In: Induced Mutations and Molecular Techniques for Crop Improvement. IAEA, Vienna. pp.347-343

Foy, C.D., 1983. Plant adaptation to mineral stress in problem soil. Iowa State J.Res. **57**(4): 339-354

Foy, C.D., 1988. Plant adaptation to acid, aluminium-toxic soils. Comm.Soil Sci.Plant Anal. **19**(7-12): 959-987

Haug, A., 1984. Molecular aspects of aluminium toxicity. Crit.Rev.Plant Sci. **1**: 345-373

Jende-Strid, B., 1993. Genetic control of flavonoid biosynthesis in barley. Hereditas. **119**: 187-204

Kaushik, R.P. and G.S. Khush, 1991. Endosperm mutants in rice: gene expression in *japonica* and *indica* backgrounds. Cereal Chem. **68**: 487-491

Khatiwada, S.P., D. Senadhira, A.L. Carpena, R.S. Zeigler, and P.G. Fernandez, 1996. Variability and genetics of tolerance for aluminium toxicity in rice (*Oryza sativa* L.). Theor.Appl.Genet. **93**: 738-744

Larsen, P.B., J. Degenhardt, C.Y. Tai, L.M. Stenzler, S.H. Howell, and L.V. Kochian, 1998. Aluminium-resistant *Arabidopsis* mutants that exhibit altered patterns of aluminium accumulation and organic acid release from roots. Plant Physiol. **117**: 9-18

Luo, M.-C. and J. Dvorak, 1996. Molecular mapping of an aluminium tolerance locus on chromosome 4D of Chinese Spring wheat. Euphytica. **91**: 31-35

Maluszynska, J. and M. Maluszynski, 1983. MNUA and MH mutagenic effect after double treatment of barley seeds in different germination periods. Acta Biologica. **11**: 238-248

Maluszynski, M. (1999) Crop germplasm enhancement through mutation techniques. In: Proc. Int. Symp. on Rice Germplasm Evaluation and Enhancement. Rutger, J.N., J.F. Robinson and R.H. Dilday (Eds.) Arkansas Agricultural Experiment Station, Fayetteville, Arkansas. pp.74-82

Maluszynski, M., B.S. Ahloowalia, A. Ashri, K. Nichterlein, and L. van Zanten, 1999. Induced mutations in rice breeding and germplasm enhancement. In: Proceedings of the 19th Session of the International Rice Commission. FAO, Rome. pp.194-204

Maluszynski, M., B.S. Ahloowalia, and B. Sigurbjörnsson, 1995a. Application of *in vivo* and *in vitro* mutation techniques for crop improvement. Euphytica. **85**(1-3): 303-315

Maluszynski, M., I. Szarejko, R. Madajewski, A. Fuglewicz, and M. Kucharska, 1988. Semi-dwarf mutants and heterosis in barley. I. The use of barley sd-mutants for hybrid breeding. In: Semi-Dwarf Cereal Mutants and Their Use in Cross-Breeding III. IAEA-TECDOC 455, Vienna. pp.193-206

Maluszynski, M., L. van Zanten, A. Ashri, H. Brunner, B.S. Ahloowalia, F.J. Zapata, and E. Weck, 1995b. Mutation techniques in plant breeding. In: Induced Mutations and Molecular Techniques for Crop Improvement. IAEA, Vienna. pp.489-504

Minella, E. and M. E. Sorrells, 1992. Aluminium tolerance in barley: Genetic relationships among genotypes of diverse origin. Crop Sci. **32**: 593-598

Polle, E., C.F. Konzak, and J.A. Kittrick, 1978. Visual detection of aluminium tolerance levels in wheat by hemetoxylin staining of seedling roots. Crop Sci. **18**: 824-827

Reid, D.A., 1971. Genetic control of reaction to aluminium in winter barley. In: Barley Genetics II. Proc. Sec. Int. Barley Genet. Symp. Nilan, R.A. (Ed.) Wash. State Univ., Pullman, WA. pp.409-413

Satoh, H., L.Q. Qu, T. Kumamaru, and M. Ogawa, 1997. Glutelin mutants induced by MNU treatment in rice. RGN. **14**: 81-84

Szarejko, I. and M. Maluszynski, 1999. High frequency of mutations after mutagenic treatment of barley seeds with Na N_3 and MNH with application of interincubation germination period. MBNL. **44**: 28-30

Ullrich, S. E. and A. Aydin, 1988. Mutation breeding for semidwarfism in barley. In: Semi-dwarf Cereal Mutants and Their Use in Cross Breeding III. IAEA-TECDOC 455, Vienna. pp.135-144

Warner, R.L., C.J. Lin, and A. Kleinhofs, 1977. Nitrate reductase-deficient mutants in barley. Nature. **269**: 406-407

Wright, R.J., 1989. Soil aluminium toxicity and plant growth. Comm.Soil Sci.Plant Anal. **20**(15/16): 1479-1497.

Anthocyanin mutations improving tomato and pepper tolerance to adverse climatic conditions

B. Atanassova, S. Daskalov, L. Shtereva and E. Balatcheva
Institute of Genetics "Prof. D. Kostov", Bulgarian Academy of Sciences, Sofia 1113, Bulgaria

Key words: anthocyaninless, pepper, plantlet, seed germination, stress conditions, tomato

Summary

The effect of four mutated genes affecting anthocyanin biosynthesis was investigated relative to tomato and pepper germination capacities and early seedling and plantlets growth. The study was performed on isogenic/near isogenic lines (IL/NIL) differing for genes *ah* (*Hoffmann's anthocyaninless*), *aw* (*anthocyanin without*) and *bls (baby leaf syndrome)* in tomato and al_1 *(anthocyaninless - 1)* in pepper. Germination responses of the IL/NILs showed that genes *ah, aw, bls* and in tomato and al_1 in pepper enhanced germination abilities under stress conditions regardless of the inherent seed properties that imparted rapid germination. This evaluation was conducted with four very different types of stress: 13°C, 33°C, 120 mM NaCl and 15% PEG-6000 water solutions. The four genes exercised no effect on root and hypocotyl elongation under the same stress conditions and their effect on plantlet growth varied depending on the genotype and the treatment. The usefulness of genes *ah, aw* and *bls* in breeding tomato cultivars tolerant to adverse climatic conditions at germination, when used as a morphological marker, is discussed.

M. Maluszynski and K.J. Kasha (eds.),
Mutations, In Vitro and Molecular Techniques for Environmentally Sustainable Crop Improvement, 139–149.

Introduction

Tomato and pepper are, as are many other cultivated species, particularly sensitive to stress conditions at germination and early onthogenic stages(Mancinelli *et al.*, 1967; Thompson, 1974; Scott and Jones, 1985; Liptay and Tan, 1985; Scott and Jones, 1986; Jones, 1986; Atanassova *et al.*, 1993; Liptay and Zariffa, 1993). A high level of uniform germination is the basis for producing vigorous plants at a uniform stage of growth. It was found that the majority of the most rapidly germinating lines from nearly 400 tomato (*Lycopersicon esculentum* Mill.) and 100 pepper (*Capsicum annuum* L.) accessions carried well-known mutated genes preventing anthocyanin production. Previous analysis of the germination responses in *aa (anthocyanin absent)*, *ae (entirely anthocyaninless)*, *ag (anthocyanin gainer)*, *ah (Hoffmann's anthocyaninless)*, *aw (anthocyanin without)* and *bls (baby lea syndrome)* anthocyaninless lines in tomato showed that only the last three genes enhanced germination. Genes *aa* and *ag* had no effect on tomato germination capacities and the effect of gene *ae* varied depending on the genotype (Atanassova *et al.*, 1995). Observations in the fields and greenhouse showed also that cotyledons in some anthocyaninless lines emerged earlier than those of their anthocyanin-containing IL/NILs.

A number of mutations that alter the intensity and the distribution of anthocyanin pigmentation or result in complete absence of anthocyanin throughout the tomato and pepper developmental stages have been identified (Rick and Chetelat, 1993; Daskalov and Poulos, 1994) and genetic studies on the anthocyanin pathway in tomato have been initiated (Goldsbrough *et al.*, 1994; Yoder *et al.*, 1994).

The branched flavonoid biosynthesis pathway produces a variety of compounds that display various functions (Mol *et al.*, 1996) and numerous benefits concerning plant tolerance to biotic and abiotic stresses have been associated with some of these compounds (Ulrichova and Sosnova, 1970; Hashimoto and Tajima, 1980; Ebel *et al.*, 1984; Zornoza and Esteban, 1984; Li *et al.*, 1993; Koes *et al.*, 1994; Bongue-Bartelsman and Phillips, 1995)

Being aware that the establishment of morphological markers correlated with tolerance to stress at any plant growth stage would be useful for the breeding of genotypes tolerant to adverse climatic conditions, experiments were designed aimed at: a) the evaluation of the effect of mutations affecting anthocyanin biosynthesis on the germination abilities of tomato and pepper under optimal and stress conditions; b) the evaluation of the effect of these genes on tomato and pepper development under stress conditions during the subsequent early growth stages (radicle protrusion to cotyledon emergence and cotyledon emergence to third leaf); c) the elucidation of the mechanisms of this effect; d) the evaluation of the usefulness of these traits in breeding programs.

Materials and methods

Plant material

The investigation was performed on nine tomato lines (Ailsa Craig, Apedice, Apeca, Monfavet 167, Monfavet 168, Piernita, Por, VF 36 and VFNT cherry) and one pepper line (Zlaten medal) and the isogenic/near isogenic lines (IL/NILs) from them differing for the following mutated genes: *ah* (*Hoffmann's anthocyaninless*), *aw* (*anthocyanin without*) and *bls (baby leaf syndrome*) in tomato, and al_1 *(anthocyaninless - 1)* in pepper. The mutant lines were obtained as a result of spontaneous or induced mutations, or by backcrossing (Maxon Smith & Ritchie,

1982; Philouze, 1991; Rick & Chetelat, 1993; Daskalov, 1980; Atanassova *et al.*, 1997a, 1997b) (Maxon Smith & Ritchie, 1982; Philouze, 1991; Rick & Chetelat, 1993; Daskalov, 1980; Atanassova *et al.*, 1997a, 1997b). The pepper line Alfi al_1 was obtained as result of hybridization between the variety Phytostop and Albena, followed by subsequent backcrosses with Albena. All experiments were carried out using seeds produced in the same year and under the same climatic conditions. Each experiment was repeated at least three times using seeds produced during three different years.

Germination conditions

Fifty seeds in each of four replicates were germinated in darkness in 14 cm Petri dishes layered with 0.5 cm thick synthetic sponge and sheet of filter paper, soaked in 25 ml distilled water, PEG-6000 or NaCl water solution depending on the treatment. Germination responses were visually scored as radicle protrusion and the criterion "time to 50% germination" was used for the assessment of the lines performance.

Seedling growth and measurements

The study of the effect of genes *ah, aw* and *bls* in tomato and al_1 in pepper on seedling growth was carried out in two experiments: 1) stress application at the stage "radicle protrusion - cotyledons emergence" only, i.e., germination up to radicle protrusion stage at 25°C and subsequent development under stress conditions; and 2) initiation of the stress treatment immediately after seed inhibition and continued up to cotyledon emergence. A total of 20 germinated seeds were used in the first experiment, or 20 dry seeds in the second experiment. These were planted in transparent growth columns that allowed direct visual observation of their further morphological development (Muthiah *et al.*, 1994). For the development of the seeds enough distilled water (or NaCl, or PEG-6000 solution, depending upon the treatment) was added to soak the filter paper at the bottom of the growth columns and to allow it to diffuse to the germinated seeds. The experiments were conducted in the dark in four replicates of 10 seeds each. Seedling development was monitored by measuring hypocotyl and root lengths for each seedling as detailed in another paper (Orbovic and Kieu, 1996).

Plantlet growth and measurement

In order to evaluate the effect of genes *ah, aw* and *bls* in tomato and al_1 in pepper on plant performance under stress at the stage "cotyledon emergence to third leaf" stage, seeds of the tomato and pepper IL/NILs were sown on Perlite in plastic trays and placed in a dark growth chamber at 25°C. The stress treatments were initiated at the commencement of cotyledon emergence and conducted in a growth chamber (16/8 h photoperiod) using four replicates of 10 plantlets. At the end of the treatment period (third leaf stage) the plantlets were harvested and the fresh leaf, stem and root material was weighed. Plantlets were fertilized weekly with Hoagland and Arnon solution (Hoagland and Arnon, 1950).

Stress conditions

Seed germination, root, hypocotyl and plantlet growth responses were evaluated under the following treatments: a) low temperature, 13 ± 1°C; b) high temperature, 33 ± 1°C; c) salt stress, 120 mM NaCl water solution; d) osmotic stress, 15% PEG-6000 water solution. The details of establishing these stress parameters have been presented previously (Atanassova *et al.*, 1995; Atanassova et al., 1997b).

Genetic variability and statistical methods

The results were expressed as the means and standard errors calculated from replicates (Lakin, 1990). The root, leaves and stems weights data were analysed by Duncan's Multiple Range Test (Duncan, 1955).

Results

Effect of anthocyanin biosynthesis mutated genes on the germination ability of tomato and pepper seeds under optimal and stress conditions

Germination responses of the IL/NILs showed that genes *ah, aw, bls* in tomato and al_1 in pepper enhanced their seed's germination abilities under stress conditions.

Previous analysis of the germination responses in the antocyanin-containing tomato lines indicated that they represented a large scale of germination capacities under optimal conditions (Atanassova *et al.*, 1997b). Significant differences were detected also in the germination rates of these lines under stress conditions. In general, the slowly germinating lines and one of these possessing moderate germination capacity, 'VF-36', performed as sensitive under all four treatments. 'VFNT cherry' performed as highly tolerant to high temperature and PEG-6000 treatment. The lines tolerant to low temperature, 'Apeca' and 'Apedice', were classified also as tolerant to salinity (Fig. 1).

The data presented in Figure 1 and Table 1 indicated that the genes *ah*, *aw* and *bls* in tomato and al_1 in pepper enhanced the germination capacities of the lines studied as in general the anthocyaninless IL/NILs germinated more rapidly than the anthocyanin-containing ones. Significant differences in the time to 50% germination were detected between the wild type lines and their *ah*-, *aw*- or *bls*-IL/NILs in tomato and al_1-IL/NILs in pepper under the four treatments. However some exceptions have to be mentioned: no differences in germination responses were observed between the anthocyanin-containing and the anthocyaninless IL/NILs differing for gene *aw* under PEG-6000 treatment and between VFNT cherry and its *aw*- and *bls*- IL/NILs under high temperature treatment.

Table 1. Germination responses (A) and plantlets growth responses (B) of pepper IL/NILs under stress conditions

Genotypes	A. Time to 50% germination (days)							
	13°C		33°C		120 mm NaCl		15% PEG-6000	
Zlaten medal	41.2 ± 5.7		7.8 ± 1.1		11.3 ± 2.5		12.0 ± 2.0	
Alfi *al 1*	28.4 ± 2.9		2.4 ± 0.4		5.8 ± 1.3		5.5 ± 1.2	
Albena *al 1*	31.4 ± 2.2		6.3 ± 1.0		9.2 ± 1.8		8.2 ± 1.6	
	B. Plantlets growth (mean fresh weight/plant in mg)							
	Leaves and stems	Root	Leaves and stems	Root	Leaves and stems	Root	Leaves and stems	Root
Zlaten medal	253 ± 28	21 ± 3	411 ± 37	39 ± 5	321 ± 20	42 ± 7	370 ± 26	39 ± 4
Alfi *al 1*	281 ± 11	30 ± 2	444 ± 42	45 ± 7	356 ± 11	50 ± 3	365 ± 19	47 ± 4
Albena *al 1*	265 ± 19	26 ± 3	450 ± 19	45 ± 4	343 ± 28	47 ± 4	392 ± 33	48 ± 6

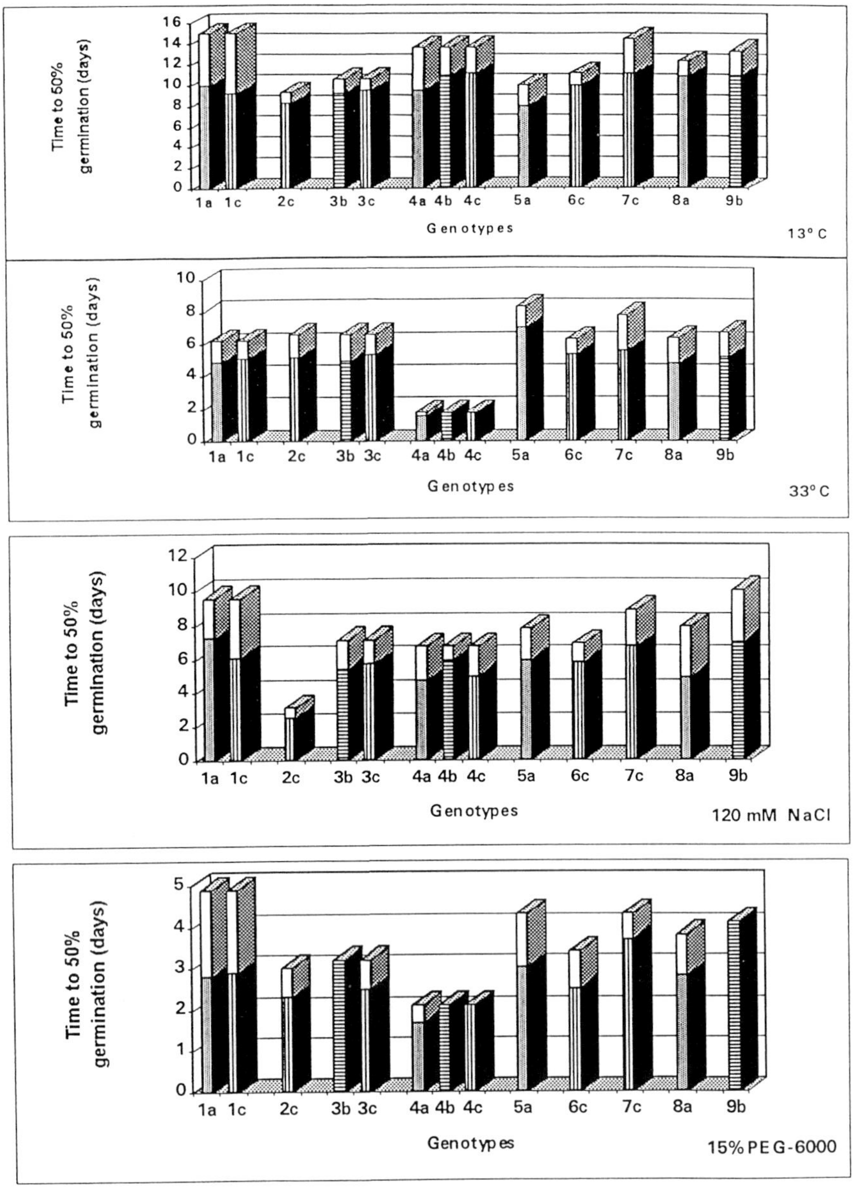

Figure 1. Germination responses of tomato IL/NILs under stress conditions. Each column presents the time to 50% germination of the anthocyanin-containing line, its lower part showing the time to 50% germination of the anthocyaninless isogenic line. Thus, the upper part of each column shows the difference in germination responses between the anthocyanin containing and the anthocyaninless IL/NIL. 1. Ailsa Craig; 2. Apeca; 3. Apedice; 4. VFNT cherry; 5. Piernita; 6. Monfavet 168; 7. Monfavet 167; 8. Por ; 9. VF 36; a -*ah*-IL/NIL; b - *aw*-IL/NIL; c - *bls*-IL/NIL

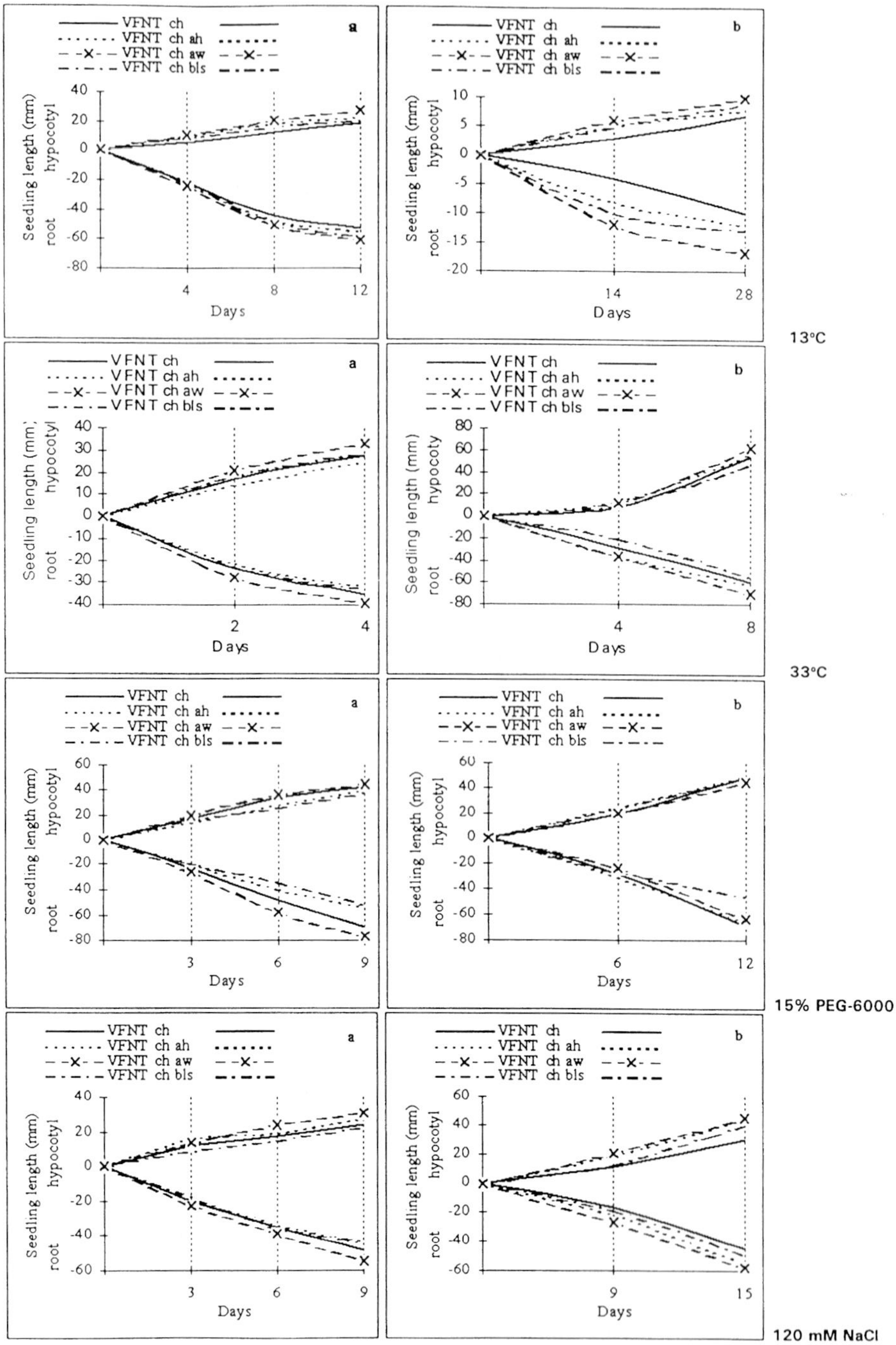

Figure 2. Growth curves for hypocotyl and root elongation of seedlings of tomato IL/NILs grown under stress conditions. a - stress application at the stage "radicle protrusion - cotyledons emergence" only; b - initiation of the stress treatment immediately after seed imbibition up to cotyledons emergence

Effect of genes ah, aw and bls on early seedling and plantlet development under stress conditions

Early seedling growth

In order to avoid repetition, and as the reactions of the ten genotypes and their IL/NILs were in general similar, seedlings growth responses will be presented and discussed only for the VFNT cherry genotype (Figure 2). The growth curves presented (Figure 2b) indicate more rapid hypocotyl and root elongation in the anthocyaninless IL/NILs under low temperature and NaCl treatments when the stress was initiated immediately after seeds imbibition. No significant differences in root and hypocotyl elongation were observed between the anthocyanin-containing line and its anthocyaninless IL/NILs under 33°C and PEG-6000 treatments.

When the stress was initiated at the stage of radicle protrusion in the seeds (Fig 2a), no differences in hypocotyl length were observed between the wild type line and its *ah*-, *aw*- and *bls* IL/NILs, regardless of the treatments. No differences in root growth were observed between the anthocyanin-containing line and its IL/NILs under three of the treatments; under osmotic stress, root elongation in the wild type line was even more intensive than that of VFNT cherry *ah* and VFNT cherry *bls*.

Plantlet development

The means and standard deviations (SD) for fresh weight/plant for three anthocyanin containing lines in tomato and one line in pepper and their anthocyaninless IL/NILs under stress conditions are presented in Table 1 and Table 2. It can be seen that under the stress

Table 2. Plantlets growth responses (min fresh weight/plant) of tomato IL/NILS to stress conditions. The Duncan test was applied separately for each group of IL/NILs and for each treatment

	13°C		33°C		120mM NaCl		15% PEG-6000	
Genotypes	leaves and stem	root	leaves and stem	root	leaves and stem	root	leaves and stem	root
A. Craig	225 ± 14^{a}	40 ± 1^{a}	285 ± 7^{ab}	18 ± 7^{a}	360 ± 8^{a}	78 ± 11^{a}	139 ± 6^{a}	48 ± 3^{a}
A. Craig *ah*	260 ± 8^{b}	40 ± 4^{a}	270 ± 12^{a}	21 ± 2^{a}	378 ± 17^{a}	99 ± 3^{ab}	127 ± 4^{a}	43 ± 6^{a}
A. Craig *bls*	210 ± 6^{a}	36 ± 2^{a}	254 ± 8^{a}	16 ± 3^{a}	392 ± 5^{ab}	94 ± 9^{a}	148 ± 7^{a}	39 ± 5^{a}
Apedice	162 ± 18^{a}	36 ± 5^{a}	245 ± 9^{a}	28 ± 6^{a}	200 ± 15^{a}	73 ± 9^{ab}	166 ± 9^{a}	68 ± 7^{a}
Apedice *aw*	165 ± 9^{a}	40 ± 1^{a}	283 ± 11^{b}	28 ± 1^{a}	192 ± 8^{a}	54 ± 6^{a}	155 ± 5^{a}	58 ± 7^{a}
Apedice *bls*	220 ± 11^{b}	52 ± 7^{b}	257 ± 5^{a}	28 ± 3^{a}	275 ± 21^{b}	71 ± 7^{ab}	231 ± 16^{b}	99 ± 12^{b}
VFNT ch.	165 ± 9^{ab}	30 ± 3^{a}	195 ± 19^{ab}	20 ± 1^{a}	236 ± 7^{ab}	65 ± 4^{ab}	180 ± 8^{ab}	65 ± 4^{b}
VFNT ch. *ah*	144 ± 16^{a}	22 ± 5^{a}	182 ± 3^{a}	18 ± 1^{a}	203 ± 11^{a}	47 ± 8^{a}	160 ± 11^{a}	45 ± 6^{a}
VFNT ch. *aw*	183 ± 7^{b}	30 ± 5^{a}	296 ± 22^{c}	25 ± 6^{a}	265 ± 6^{c}	89 ± 10^{c}	206 ± 13^{b}	91 ± 9^{c}
VFNT ch. *bls*	155 ± 6^{a}	27 ± 2^{a}	168 ± 6^{a}	16 ± 2^{a}	196 ± 9^{a}	52 ± 2^{a}	165 ± 2^{a}	47 ± 8^{a}

treatments, no uniformity in the anthocyaninless plantlet responses occurred compared to those of their anthocyanin-containing IL/NILs. For example, under low temperature stress Apedice *bls* was more tolerant than Apedice, while no significant differences in root and stem plus leaves fresh weights were detected between 'Ailsa Craig' and VFNT cherry and the *bls*-anthocyaninless IL/NILs. Nevertheless it must be noted that in general there was no negative effect of the three anthocyaninless genes on plantlet development under the treatments tested.

Discussion

This study has revealed an enhancing effect of three anthocyaninless mutated genes on tomato and one such mutant on pepper germination abilities under stress conditions. Our attempts to determine the seed characteristics that might be affected or altered by genes *ah*, *aw* and *bls* in tomato provided evidence that the seed coat and endosperm have been affected in a way that is advantageous for more rapid germination, although this effect might not be limited to these two seed components (Atanassova *et al.*, 1997a). The relatively significant number of anthocyanin containing lines studied (especially in tomato), as well as the differences detected in their germination responses under non-stress and stress conditions, indicated that genes *ah*, *aw* and *bls* in tomato and al_1 in pepper enhanced germination (except for gene *aw* under PEG-6000 treatment) regardless of the inherent germination abilities of the line under optimal, low and high temperature, osmotic and salinity stresses.

The exceptions observed concerned mainly the rapidly germinating tomato line VFNT cherry when germinated under high temperature and 15% PEG-6000 water solution. At least two characteristics may have contributed to the rapid germination of this line under the above-mentioned treatments: its inherent tolerance to these stresses and; probably some inherent seed property advantageous for germination. Theoretically, the anthocyaninless VFNT cherry IL/NILs possessed the same characteristics. As the germination capacity of one line had an upper limit it might be suggested that line VFNT cherry had attained its maximum germination potential under the treatments tested and the effect of genes *ah*, *aw* and *bls* could not add to those of the two above mentioned characteristics. Based on the presented experiments, there were no significant differences in germination responses between Apedice, VF 36, VFNT cherry and their *aw*-IL/NILs under somatic stress. Having in mind the differences in the molecular characteristics of the three genes (Goldsbrough *et al.*, 1994; Yoder *et al.*, 1994), it might be suggested that each one of them could affect different seed properties and/or metabolic process/es, or the same seed properties and/or metabolic process in different way, despite their similar effect on germination.

The analysis of the presented data indicated that because of the effect of genes *ah*, *aw* and *bls*, the anthocyaninless IL/NILs of the slowly germinating and highly sensitive to stress lines like 'Monfavet 167', VF 36 and Ailsa Craig possessed germination potential more or less equal to that of the rapidly germinating and highly tolerant lines VFNT cherry, Apeca and Apedice. It is known that the shorter time necessary for the germination of 50% of the seeds is correlated with more tolerance to the given stress at germination. However, in the rapidly germinating genotypes the shorter time to 50% germination under stress might be due to some inherent property/s of the seed that lead to rapid germination regardless of the conditions, and not to tolerance to stress *per se* (Scott and Jones, 1985; Jones, 1986). Nevertheless, despite the different mechanisms for overcoming the stress, the performance of the rapidly germinating genotypes is a kind of tolerance to stress at germination. In this sense, the effect of the three genes might be characterized as increasing the potential of the genotype to germinate rapidly and to overcome stress at germination.

In general, no significant differences in root and hypocotyl growth between the anthocyanin-containing and the anthocyaninless IL/NILs were observed when stress was applied at the stage "radicle protrusion to cotyledons emergence". This indicates that the enhancing effect of genes *ah*, *aw* and *bls* in tomato and al_1 in pepper was probably limited to the stage of radicle protrusion and did not affect the subsequent developmental stage. On the other hand, the significantly more intensive development of the anthocyaninless seedlings under cold and salinity stress observed (when stress treatment was initiated at the moment of seed imbibition), indicated that earlier cotyledon emergence in the anthocyaninless tomato and pepper lines observed under greenhouse and fields conditions was probably due to the earlier germination of these genotypes and not to some enhancing effect of genes *ah*, *aw*, *bls* and al_1 on hypocotyl and root growth. This theory is supported by the fact that no differences in seedling growth between the anthocyaninless and the anthocyanin-containing lines were detected under high temperature and osmotic stress. As mentioned previously, the wild type line was highly tolerant to these stresses at germination and genes *ah, aw* and *bls* enhanced slightly or not at all its germination abilities. On the other hand VFNT cherry performed was highly sensitive to low temperature and salinity stress. Under these treatments, the enhancing effect of the three genes was significantly more important, that resulted in earlier germination, subsequent better root and hypocotyl growth and cotyledon emergence. The diversity in the anthocyaninless plantlet growth responses under stress when compared to those of the anthocyanin-containing IL/NILs from them might suggest that their reactions depended on the interaction genotype x anthocyaninless mutated gene x nature of the stress. The results obtained provide evidence that supplementary experiments (already under way), including larger numbers of genotypes and more intensive stress treatments are necessary to get clear information on the effect of genes *ah, aw* and *bls* in tomato and al_1 in pepper on plant development under stress at the stage "cotyledons emergence to third leaf".

To be able to survive a stress, that imbibed seeds must either be quite insensitive to this stress, or else they must possess some mechanism for adjustment. The analysis of the anthocyanin-containing lines germination responses under stress treatment indicated that the line VFNT cherry, highly tolerant to high temperature and osmotic stress, performed as highly sensitive to low temperature and salt treatment. Conversey, Apeca, tolerant to low temperature and salt stress was sensitive to high temperature. Thus, the capability of one accession to cope with one or more types of stress might not always be useful for coping with another type of stress. On the other hand, stresses in nature can occur simultaneously or sequentially and may counteract or intensify each other's effects. Some of them might be envisaged and hence controlled or avoided, but for others, like the low or high temperature stresses, it is impossible to do so. Thus, in case of low temperature stress, for example, many of the seeds of one directly seeded cultivar, tolerant to high temperature at germination may not germinate, or would germinate so irregularly that plants would grow differentially and crop development would be extremely variable and delayed. In this sense the development of tomato and pepper cultivars with an increased germination potential, conditioned by genes *ah, aw* or *bls* in tomato and al1 in pepper, being expressed regardless of the nature and the intensity of the stress, or the intrinsic capability of an individual accession to cope with a given type of stress, would contribute significantly to the efficiency of uniform stand establishment. Moreover, in general, no negative effect of the four genes on subsequent seedlings and plantlet growth was detected. On the contrary, on the basis of comparative field trials it was found that the pepper variety Albena exceeded Zlaten medal in early and total yield, (Daskalov, 1975) and Apedice exceeded Apedice *aw* in early yield, (unpublished data). Real possibilities exist for the characterization of genes *ah* and *bls* (as gene *aw* was recently characterized) (Goldsbrough *et al.*, 1994), might make their transfer by means of molecular techniques and usefulness in plant breeding more rapid and efficient. And last, but perhaps the most important, is the fact that because of the maternal effects related to the germination rate when conditioned by genes *ah, aw* and *bls* (Atanassova *et al.*, 1997a), the phenomenon can be used also in the breeding of tomato hybrid varieties. A number of tomato hybrids (Balkan, Kom, Rila, Dorina, Precoce, Kamena) developed at the Institute of Genetics, Sofia and by Geosem Select, Sofia possess seed parents that carry one of the tree genes (*ah, aw* or *bls*). These hybrids, as well as the anthocyaninless pepper variety Albena, are widely used in practice.

Conclusions

Our main findings in this study might be summed up as follows:

1. Genes *ah, aw* and *bls* in tomato and al_1 in pepper enhanced germination in tomato and pepper lines regardless of their inherent ability to germinate under optimal and stress conditions and/or the inherent seed properties that imparted rapid germination;
2. The four genes exercised a kind of universal effect on tomato and pepper germination capacities as they increased the germination potential of the individual accession under optimal conditions and under a relatively large scale of stresses different in nature;
3. Genes *ah, aw* and *bls* in tomato and al_1 in pepper exercised no enhancing effect on early seedling growth and plantlets develop under 13°C, 33°C, 120 mM NaCl and 15% PEG-6000 water solutions.

The breeding of tomato and pepper cultivars tolerant to adverse climatic conditions is a long and difficult process mainly because of the polygenic control of the character and the stage specificity of the genes involved. Thus the establishment of a morphological marker correlated with tolerance to stress even at one plant growth stage, is meaningful and significant.

References

Atanassova, B., L. Shtereva, and S. Daskalov, 1995. Effect of mutant genes on the germination abilities of tomato and pepper under optimal and stress conditions. In: Induced Mutations and Molecular Techniques for Crop Improvement. IAEA, STI/PUB/972, Vienna. pp.643-645

Atanassova, B., L. Shtereva, and E. Molle, 1993. New potential for the improvement of tomato salt tolerance during germination. In: Proc. of the XIIth Eucarpia Meeting on Tomato Genetics and Breeding. Stamova L. (Ed.). Maritza Veg. Crops Res. Inst., Plovdiv. pp.81-87

Atanassova, B., L. Shtereva, and E. Molle, 1997a. ffect of three anthocyaninless genes on germination in tomato (*Lycopersicon esculentum* Mill.). I. Seed germination under optimal conditions. Euphytica. **95**: 89-98

Atanassova, B., L. Shtereva, and E. Molle, 1997b. Effect of three anthocyaninless genes on germination in tomato (*Lycopersicon esculentum* Mill.). II. Seed germination under stress conditions. Euphytica. **97**: 31-38

Bongue-Bartelsman, M. and D.A. Phillips, 1995. Nitrogen stressregulates gene expression of enzymes in the flavonoid biosynthetic pathway of tomato. Plant Physiol. Biochem. **32**(5): 539-546

Daskalov, S. and J.M. Poulos, 1994. Update *Capsicum* gene list. Capsicum and Eggplant Newsl. **13**: 15-26

Duncan, D.B., 1955. Multiple range and multiple F test. Biometrics. **11**: 1-21

Ebel, J., W.E. Schmidt, and R. Loyal, 1984. Phytoalexin synthesis in soybean cells; Elicitor induction of phenylamine ammonia-lyase and chalcone synthase mRNA and correlation with phytoalexin accumulation. Arch. Biochem. Biophys. **232**: 240-248

Goldsbrough, A., F. Belzile, and J.I. Yoder, 1994. Complementation of the tomato anthocyanin without (*aw*) mutant using the dehydroflavonol 4-reductase gene. Plant Physiol. **105**: 491-496

Hashimoto, T. and M. Tajima, 1980. Effects of ultraviolet irradiation on growth and pigmentation in seedlings. Plant Cell Physiol. **21**: 1559-1571

Hoagland, D.R. and D. Arnon, 1950. The water culture method for growing plants without soil. California Agric. Exp. Sta. Circular. **347**: 1-32

Jones, R.A., 1986. High salt tolerance potential in *Lycopersicon* species during germination. Euphytica. **35**: 575-582

Koes, E., F. Quattrocchio, and J.N.M. Mol, 1994. The flavonoid biosynthetic pathway in plants: function and evolution. BioEssays. **16**: 123-132

Lakin, G.F. (1990) Biometrics. Vishaia shkola (In Russian), Moscow

Li, J., T.-M. Ou-Lee, R. Raba, R.G. Amundson, and R.L. Last, 1993. *Arabidopsis* flavonoid mutants are hypersensitive to UV-B irradiation. The Plant Cell. **5**: 171-179

Liptay, A. and C.S. Tan, 1985. Effect of available water on germination of polyethylene glycol (PEG) pretreated tomato seeds. J. Amer. Soc. Hort. Sci. **110**: 748-751

Liptay, A. and N. Zariffa, 1993. Testing the morphological aspects of polyethylene glycol primed tomato seeds with proportional odds analysis. Hort. Sci. **28**(9): 881-883

Mancinelli, A.L., Z. Yaniv, and P. Smith, 1967. Phytochrome and seed germination. I.Temperature dependence and relative P_{fr} levels in the germination of dark-germinating tomato seeds. Plant Physiol. **42**: 333-337

Mol, J., G. Jenkins, E. Schafer, and D. Weiss, 1996. Signal perception, transduction and gene expression involved in anthocyanin biosynthesis. Crit. Rev. Plant Sci. **15**(5&6): 525-557

Muthiah, S., E.L. David, and W.M. Harris, 1994. Staging soybean seedling growth from germination to emergence. Crop Sci. **34**: 289-292

Orbovic, V. and K. Kieu, 1996. A semi-automatic method for measurement of seedling length. Plant Growth Regul. **20**: 303-305

Rick, C.M. and R.T. Chetelat, 1993. Stocks Lists. Tomato Genet. Corp. **43**: 53-87

Scott, S.J. and R.A. Jones, 1985. Cold tolerance in tomato. I. Seed germination and early seedling growth of *Lycopersicon ecsulentum*. Physiol. Plant. **65**: 487-492

Scott, S.J. and R.A. Jones, 1986. Cold tolerance in tomato II. Early seedling growth of *Lycopersicon* ssp. Physiol. Plant. **66**: 659-663

Thompson, P.A., 1974. Characterization of the germination responses to temperature of vegetable seeds. I. Tomatoes. Sci Hortic. **2**: 35-54

Ulrichova, M. and V. Sosnova, 1970. Effect of phosphorous deficiency and anthocyanin content in tomato plants. Biol. Plant. **12**: 231-235

Yoder, J.I., F. Belzile, Y. Tong, and A. Goldsbrough, 1994. Visual markers for tomato derived from the anthocyanin biosynthetic pathway. Euphytica. **79**: 163-167

Zornoza, P. and R.M. Esteban, 1984. Flavonoid content of tomato plants for the study of nutritional status. Plant Soil. **82**: 269-271.

The use of somaclonal variation and *in vitro* selection for improvement of plants to environmental stresses

J. Bouharmont, H. Batoko, P. Bertin, C. Costa de Macedo, S. Lutts, N. Skali Senhaji, V. Van Sint Jan and G. Zhu
Laboratoire de Cytogénétique, Université Catholique de Louvain, Belgium

Keywords: Oryza sativa, rice, somaclonal variation, *in vitro* selection, cold tolerance, disease resistance, hybrid vigour

Summary

The programme was established for the development of rice culture at high altitude. Somaclonal variations and *in vitro* selection were proposed for improving cold tolerance and resistance to a disease, bacterial brown sheath rot due to *Pseudomonas fuscovaginae*, which is prevalent in cool and humid conditions of tropical mountains. The rice varieties used for investigations and *in vitro* cultures were obtained from IRRI germplasm and from regional or national breeding programs, mainly from West Africa, Burundi and Madagascar. Many different *in vitro* selection procedures were used. In general, more drastic treatments reduced the number of surviving calli and regenerated plants and improved the frequency of aberrant phenotypes, but the consequences of the selection were different according to the parent varieties. Application of strong selection pressures for abiotic stresses, reduce drastically the number of regenerated plants while lower pressures are not very efficient. Many plants and progenies resistant to the different constraints have been found without application of any *in vitro* selection and they can be screened in large populations on the field or in controlled conditions. The experimental results confirmed the frequent occurrence of variations in plants derived from callus culture in rice and the broader variability in the regenerated populations. The new characters were generally transmitted to the next generations. Somaclonal mutations involved many morphological, agronomical, physiological and biochemical modifications in comparison to the parent varieties.

M. Maluszynski and K.J. Kasha (eds.),
Mutations, In Vitro and Molecular Techniques for Environmentally Sustainable Crop Improvement, 151–159.

Introduction

Induction of stable mutations in plants regenerated from tissue cultures has been reported in several crop species. The frequency of these somaclonal variants is particularly high in rice and could be sufficient for application in plant breeding as well as for the study of the mechanisms of stress resistance.

Our program started in the frame of a European research project (STD) for the development of rice culture at high altitude. Somaclonal variations and *in vitro* selection were proposed for improving cold tolerance and resistance to bacterial brown sheath rot due to *Pseudomonas fuscovaginae*, which is an important disease in cool and humid conditions of tropical mountains. Research was later extended to salt and aluminium tolerance, which are two major constraints for rice culture at lower altitudes, particularly in tropical Africa and South America.

Improvement of rice varieties was the long-term objective of the programme, but it was also a good opportunity for studying the constraints and the mechanisms of stress tolerance at the cell and plant levels. These more fundamental researches involved both somaclonal lines and existing cultivars and it was thus possible to compare genotypes resulting from our experimental work and from classical breeding.

Material and methods

The rice varieties used for investigations and *in vitro* cultures were obtained from the International Rice Research Institute (IRRI) germplasm and from regional or national breeding programs, mainly from West Africa, Burundi and Madagascar.

Callus development was induced from mature embryos, and plantlets (R_0) were regenerated after a variable number of subcultures. Many different *in vitro* selection procedures were used: the calli were stressed for time durations of a few weeks to six months (sometimes only 30 min for disease resistance); during callus proliferation and/or on the regeneration medium; once or several times, with very strong or mild selection pressures (different temperatures or salt and aluminium concentrations). In general, more drastic treatments reduced the number of surviving calli and regenerated plants and improved the frequency of aberrant phenotypes, but the consequences of the selection were different according to the parent varieties. In fact, tolerant plants have been found as well in the progeny of unselected calli as well as after very different treatments.

More or less extensive field trials were performed in several African countries: Burundi for cold tolerance and disease resistance; Togo for cold tolerance and modification of various agronomical traits; Senegal for salt or aluminium tolerance. More precise tests and crossings were performed in a phytotron or greenhouse and many laboratory analytical techniques were used for the study of the tolerance mechanisms.

Results and discussion

Chilling tolerance

Calli were obtained from mature embryos of two landraces and of two new varieties bred in Burundi, characterized by different levels of cold tolerance. They were submitted to different cold stresses and a large number of plants were regenerated. The efficiency of *in vitro* selection procedures was dependent on the variety and thus it was necessary to adapt the treatment to the genotype (Bertin *et al.*, 1995). Some R_0 plantlets were tested for cold tolerance in a phytotron at different temperatures and for various durations according to the tolerance of the parent genotypes, generally at 10°C for 10 days. Other plants were cultivated in the field at different altitudes in Burundi: the selection was most effective at 1650 m. Chilling tolerance was also evaluated by measuring electrolyte leakage from stressed leaf fragments and chlorophyll fluorescence.

Regenerated plants and their progenies generally appeared more variable than the parent genotypes for morphological traits and grain production; natural selection pressures applied in the field during the first generations resulted in improved somaclonal families (Bertin *et al.*, 1996a). The effectiveness of *in vitro* selection applied on calli depended on the parent genotype: in the most sensitive cultivar, the cold stress did not result in improved performances compared with unstressed calli; on the contrary, in a cold tolerant variety, the longest cold treatments led to the best somaclonal lines for most parameters (Bertin *et al.*, 1997a).

Modifications were also observed for physiological characters related to cold tolerance: lower electrolyte leakage, higher chlorophyll fluorescence, improved survival and seed production at low temperature (Bertin and Bouharmont, 1997). A good correlation was observed between these characters in parent cold tolerant varieties, but that was not the case for lines derived from regenerated plants. Electrolyte leakage was generally lower in somaclonal families derived from some varieties, indicating an improved chilling resistance of cell membranes, while the lower values of the ratio between variable and maximum fluorescence in cold stressed leaves indicated a lower photochemical efficiency of chloroplast photosystem II at low temperature. On the contrary, families derived from other varieties showed higher electrolyte leakage, but higher chlorophyll efficiency at low temperatures (Bertin *et al.*, 1996b; 1997b).

The fatty acid composition of lipid was analysed in leaves of several varieties and in some selected somaclonal lines. Chilling tolerance of varieties was correlated with higher double bond index and higher proportion of linolenic acid in the phospholipids and galactolipids, but not in the total lipids nor neutral lipid fraction. Altered proportions of fatty acids were observed in leaf phospholipids and/or galactolipids in somaclonal lines derived from two varieties, but not for a third one. These changes resulted generally in higher double bond index and higher proportion of linoleic and linolenic acids related either with a lower ratio of C16 to C18 fatty acids or with higher unsaturation levels of C18 fatty acids (Bertin *et al.*, 1998).

Relations between cold tolerance and esterase pattern have been reported in literature. Thus that enzyme has been analysed by PAGE electrophoresis in calli of two cold tolerant and two susceptible varieties, as well as in calli selected and unselected for cold tolerance in four other varieties. The number of isoenzymatic bands for esterase was indeed higher in both susceptible varieties. After selection for cold tolerance, some esterase bands present in the parent genotypes had disappeared from the selected calli of two varieties, while the isoenzymatic patterns remained unmodified in the other two.

The same genotypes have been used for proline analyses in stressed and unstressed calli. The proline rates were higher in cold susceptible varieties and they increased faster after cold treatment of the cultures. The proline content of the calli cultivated at low temperature (11-15°C) was always lower in selected calli than in unselected ones. Thus it is clear that proline accumulation is the result of the cell reaction to a constraint, which is more important in cold susceptible varieties or in cell strains maintained at low temperature. Taken together, the results of physiological and biochemical analyses indicate that several different mechanisms are involved in the chilling tolerance of rice varieties, while selected somaclonal families are characterized by only one or few modifications.

From a practical point of view, a number of somaclonal lines were tested for several generations in the field at various localities of Burundi. According to the last reports, some of them maintained a better chilling tolerance than their parent varieties, but the level of improvement was rather low and did not overcome the performance of some existing varieties.

Salt tolerance

Callus cultures were induced in three salt sensitive or moderately resistant cultivars received from IRRI. R_0 plants were regenerated after application of different salt stresses. These plants, as well as R_1 and R_2 fertile plants randomly chosen in their progenies were cultivated without addition of salt. For most agronomical and morphological parameters, these plants showed lower performances than the parent cultivars, but the variability was strongly increased and tissue culture-derived plants often gave the maximal values. The importance of somaclonal variation depended on the genotype, the character considered, the NaCl concentration used for *in vitro* selection and the stage of stress imposition (Lutts *et al.*, 1998).

Salt resistance has not been tested on R_0 generation, but 154 R_1 families, belonging to the three varieties, were used in a survival test in a hydroponic solution containing 30 mM NaCl. The treatment was applied to 20 day-old plantlets for 21 days: 17 families showed higher survival rates compared to their parent varieties and they were used for physiological characterization. Physiological traits observed in R_1 plants at the end of the salt resistance test were the mineral composition of roots and leaves, viability (reduction of 2, 3, 5-triphenyl tetrazolium chloride), leaf cell membrane permeability and osmotic potential.

Physiological modifications were observed in four R_1 families belonging to both varieties. The first was characterized by a better discrimination between K and Na ions in the leaves of plants cultivated in the presence of NaCl: the higher K/Na ratio was due to a lower accumulation of Na and a conservation of the K level. In a second family, the segregation of Na to old leaves and K to younger ones was significantly increased. On the base of growth rate and viability tests, the salt tolerance of the plant tissues was improved in a third progeny. Finally, the last family was characterized by a lower accumulation of Zn.

The same stress and screening tests were applied to the progeny (R_2) of the four selected R_1 families. Only the modifications observed in the first two selected families (K-Na discrimination between roots and shoots and Na accumulation in older leaves) have been transmitted to R_2 plants.

During a more limited experiment of *in vitro* selection, 15 families derived from a single variety ('I Kong Pao') were tested in the field in Senegal. Plants had been regenerated from calli after *in vitro* selection; some of them had been cultivated in hydroponic conditions with NaCl; other ones were cultivated in soil and selection for salt tolerance was performed on R_1 plants. The progenies of these plants were irrigated using fresh or salt water (conductivity of 3.5 mS/cm). Thus the families tested in Senegal represented R_1 or R_2 generations. Significant differences were observed in several lines and one of them showed improved salt tolerance and could be compared to standard resistant genotypes. It was characterized by a moderate reduction of yield, biomass, fertility and yield index when cultivated in saline conditions.

In the R_2 generation of a final experiment, the family R2-1-23 exhibited higher performances than the parent variety I Kong Pao when cultivated in the presence of NaCl in relation with survival rates, flowering and number of spikelets. However, salinity resistance was not transmitted to the next generation: in spite of a higher number of spikelets per fertile plant, R3-1-23 did not perform better than the parent cultivar because of a very low seed setting. Hybrids have been obtained between that somaclonal family and the parent variety as well as with the very salt sensitive IR31785. Most F_1 hybrids cultivated in the absence of salt were more vigorous and exhibited better performances than the mid-parent, suggesting a heterotic effect on yield-related parameters. F_2 populations were screened for salinity resistance and a clear improvement was observed for yield in stress conditions for populations issued from I Kong Pao x R3-1-23 and IR31785 x R3-1-23.

Improved vigour of F_1 hybrids between I Kong Pao and R3-1-23 were unexpected and interesting: crossings between individual plants of that variety never led to heterotic effects; thus in spite of their very close relationships, the somaclonal family and its parent variety showed very clear heterosis in their F_1 and F_2 hybrids. The comparison of F_1 and F_2 was difficult because of the number of available plants and culture conditions were different. Seed production was lower in F_2 than in F_1, but these F_2 populations were still better than their parents for several agronomical characters in the absence of NaCl. When the plants were cultivated in the presence of salt (30 mM), the different F_2 hybrids showed better performances than the salt susceptible genotype, comparable to those of a tolerant variety.

In vitro selection and characterization of regenerated lines were completed by more fundamental research on salt responses of different rice genotypes. Some tall *indica* landraces appeared resistant at all developmental stages while the response was different according to the stages in some *japonica* varieties and in modern IRRI genotypes: the effect of saline conditions was more pronounced during the vegetative or reproductive phase according to the genotypes. For the vegetative phase, young seedlings were generally the most sensitive, but there were also some differences between varieties (Lutts *et al.*, 1995).

A comparison of four more or less tolerant or susceptible cultivars growing in various saline solutions showed a lower concentration of Na, Cl and Zn in roots and shoots of the tolerant ones, but a higher concentration of K in the presence of NaCl. In the same conditions, P translocation from roots to leaves was reduced in the most sensitive variety. In tolerant varieties, Na and Cl were mainly stored in oldest leaves, where the K level was lower. In the presence of NaCl, the osmotic potential of the roots and leaves was lower in salt resistant genotypes. Proline did not appear to be involved in osmotic adjustment: a lower proline accumulation in salt resistant cultivars could reflect a less important perturbation of cell metabolism (Lutts *et al.*, 1996ab) .

Several cytophysiological tests showed that NaCl induces an acceleration of senescence in all leaves of sensitive genotypes and more specifically in oldest leaves of tolerant ones (Lutts *et al.*, 1996c). NaCl induced an increase of ethylene production more important in salt tolerant cultivars. On the other hand, application of putrescine on the leaves reduced the effect of salt on growth and viability, and increased ethylene synthesis, suggesting that putrescine could be involved in salt tolerance (Lutts *et al.*, 1996d).

Aluminium tolerance

Aluminium toxicity is important in acid soils, but it is generally difficult to discriminate between the effects of aluminium itself, acidity and of mineral imbalance induced by acidity. The same problem appears also during *in vitro* selection experiments since the pH of the culture media used for screening tolerant cells must be sufficiently low for solubilization of aluminium, involving lower phosphate and calcium concentrations. Besides, an adaptation of the substrate used for *in vitro* culture is necessary, agar media being liquid at low pH.

Calli of an Al-sensitive rice variety and of an Al-tolerant one were cultivated on normal and modified (acid) media and in the presence of various concentrations of aluminium sulfate. Selection pressures have been applied once or twice during callus proliferation. Regenerated R_0 plants were then screened in hydroponic solutions for their Al-tolerance and 9 fertile plants were selected. The fact that several plants, which derived from the same embryo gave different responses to stress suggests that the mutations have appeared during the *in vitro* culture. The progenies of these 9 plants were tested during four generations: Al tolerant plants were present in three of these families. One of the families derived from a plant regenerated from an untreated callus, the other two came from calli cultivated in the presence of 250 and 1000 mM Al (Van Sint Jan *et al.*, 1997). For each generation, ten-day old plantlets were maintained for 80 days in a nutritive solution containing 1500 mmol/l Al. Under these conditions, all plants of the parent varieties died. Plants were resistant when, after the test, they remained entirely green or exhibited very limited leaf necrosis. In the three families, the proportion of resistant plants increased regularly in the successive generations, reaching 70 to 80% in R_4; nevertheless, 10 to 30% of the plants were still sensitive and it was not possible to isolate uniformly tolerant families.

Several experiments have been performed in order to study the tolerance mechanisms at the cell and plant levels. At the cell level, selected and unselected calli of these varieties have been used; selection was performed after culture on media containing aluminium sulfate; for susceptible varieties, mineral composition was analysed in some necrotic and healthy calli. In selected calli, aluminium tolerance seems to result from a lower Al uptake and from the persistence of rather high ratios between P or Ca and Al. In contrast, the calli of the standard tolerant variety used for comparison showed a higher relative concentration of Al. They seem to tolerate such a high concentration because they have a physiological mechanism working at the cell level. Four enzymes were analysed in selected calli and a modified pattern was observed for esterase in a single variety.

Similar experiments were performed on roots and leaves of R_1 plants of tolerant families cultivated in solutions containing various Al concentrations, and compared to tolerant and susceptible varieties. From the results, it appears that some tolerance mechanisms seem to be different in selected families and in the resistant varieties used for comparison. A reduction of aluminium uptake by the roots and a lower translocation to the rest of the plant seem to be particularly involved. Besides, the relative concentrations of P, Ca and Mg remain rather high, particularly in the leaves.

Haematoxylin staining of root tips after Al treatment has sometimes been proposed for screening Al resistance, the presence of a larger unstained subapical zone characterizing resistant genotypes. In fact, one of the two tolerant varieties used in an experiment showed indeed a large unstained zone, while a second one gave a staining pattern similar to that of the sensitive varieties. Such a difference confirms that plant tolerance can be due to physiological cell adaptations as well as to a lower uptake of Al by the roots.

Resistance to bacterial sheath brown rot

The bacterial disease caused by *Pseudomonas fuscovaginae* is a major constraint for rice cultivation in cold and humid highland environments. Bacterial toxins were isolated from cultures, concentrated, and their activity was tested by their toxicity for some fungi. Several components are present in the toxin and the major bioactive compound is a syringotoxin. Purified extracts were effective on the plants at all growth stages and they induced the symptoms normally observed on diseased plants, particularly a reduction of culm elongation resulting in a poor panicle exsertion and partial spikelet sterility. On the other hand, the severity of the damages induced by the toxin were related to the level of the cultivar susceptibility to the pathogen itself (Batoko *et al.*, 1994; 1997). Thus, the toxin could be used for screening susceptibility to *P. fuscovaginae* at the plant level. On the other hand, phytotoxicity of the toxin was also efficient at the cell level, inducing a high electrolyte leakage from the calli. The leakage level was correlated with the resistance of the plants to bacterial disease. Bacterial syringotoxin affects H^+-ATPase activity of rice plasma membranes by inducing permeabilization; the data suggest that the plasma membrane H^+-ATPase is a direct or indirect target for the bacterial toxin, resulting in a growth inhibition. These results justified the use of purified toxin for *in vitro* selection of resistance to *P. fuscovaginae*.

Calli of two rice varieties were used for *in vitro* selection, a disease tolerant and a susceptible one. Addition of toxin to agar medium resulted in an important necrosis of the callus areas, which were directly in contact with the medium, while the other tissues were not affected. A decrease of toxin concentration in the medium was also observed: calli of the susceptible variety induced a faster disappearance of the toxin from the medium. For improving access of toxin to the cells, small callus pieces were soaked for 8 hours or vacuum infiltrated during 30 minutes in toxin solutions before culture and regeneration on normal media. These pretreatments drastically reduced callus growth.

Plants from both cultivars were regenerated and screened for toxin sensibility at the booting stage: they showed lower mean severity rates than parent genotypes and plants regenerated from untreated calli. The R_1 and R_2 progenies of the most tolerant R_0 plants were infected by *P. fuscovaginae* and the disease severity was scored at maturity: in these tests, five families (three derived from the susceptible and two from the tolerant variety) exhibited significantly lower disease symptoms than the tolerant cultivar. The frequency of symptomless plants was also higher in these families. From these families, four derived from calli treated by vacuum infiltration and only one from soaking. The offspring of selected families have been planted in Burundi under conditions promoting the development of bacterial sheath brown rot. Unfortunately, observations have not been possible due to the political situation in the country.

Conclusions

The experimental results confirmed the frequent occurrence of variations in plants derived from callus culture in rice and the broader variability in the regenerated populations. The new characters were generally transmitted to the next generations. Somaclonal mutations involved many morphological, agronomical, physiological and biochemical modifications in comparison to the parent varieties.

Application of selection pressures on callus cultures can reduce the number of plants to be screened after regeneration. Nevertheless, at least for abiotic stresses, strong selection pressures reduce drastically the number of regenerated plants while lower pressures are not very efficient. Thus, tests must be applied on the original varieties in order to adapt the selection procedure to their susceptibility level. In fact, many plants and progenies resistant to the different constraints have been found without application of any *in vitro* selection and they can be screened in large populations on the field or in controlled conditions.

The stability of the cold and salt tolerance of somaclonal lines has been confirmed in laboratory and field trials. Nevertheless, the increasing, but partial proportion of aluminium tolerant plants in some progenies has not been explained. The tolerance levels were significantly improved, but remained rather low. It was not possible to find a new genotype better than the best available varieties.

Physiological and biochemical studies on varieties characterised by various tolerance levels showed the simultaneous presence of several adaptations at the cell or plant level. Modifications were also found in somaclonal lines selected for tolerance, but only one modified trait was generally observed in a particular line. Thus it could be expected that accumulation of various adaptations by crossing different lines or hybridization between varieties and somaclonal lines could lead to an improvement of the tolerance level. On the other hand, these somaclonal lines can be used for studying the effect of individual mutations on the physiology and on the biological behaviour of the plants. Such a study is more difficult in varieties combining several adaptations.

The role of proline in tolerance to several stresses has been frequently discussed. From our observations, it appears that proline does not represent a mechanism of resistance to abiotic stresses: on the contrary, the level of that amino acid increases in stressed susceptible genotypes, indicating a reaction of the cells submitted to a constraint.

The heterosis which was observed in F_1 and F_2 hybrids between a somaclonal line selected for salt tolerance and the parent variety was rather unexpected since the parental plants were probably genetically identical except for only very few genes. We have no satisfactory explanation, but such mutants could be useful for the study of heterosis in general.

In vitro selection for resistance to a bacterial disease was performed on a smaller scale. Thus it is difficult to compare the results of these experiments to what was observed with respect to abiotic stresses. Nevertheless, they showed the possibility of improving resistance to bacterial rot after cell selection by a toxin isolated from the bacteria. On the other hand, the toxin does not seem to go into the callus by simple diffusion; instead, the toxin is rapidly taken by the tissues and disappears from the surrounding solution. That could indicate that the disease symptoms result from a specific reaction between the syringotoxin and some cell components.

References

Batoko, H., J. Bouharmont, J.-M. Kinet, and H. Maraite, 1997. Involvement of toxins produced by *Pseudomonas fuscovaginae* in aetiology of rice bacterial sheat brown rot. J. Phytopathology. **145**: 525-531

Batoko, H., J. Bouharmont, and H. Maraite, 1994. Inhibition of rice (*Oryza sativa* L.) seedling elongation by *Pseudomonas fuscovaginae* toxin. Euphytica. **76**: 139-143

Bertin, P. and J. Bouharmont, 1997. Use of somaclonal variation and *in vitro* selection for chilling tolerance improvement in rice. Euphytica. **96**: 135-142

Bertin, P., J. Bouharmont, and J.-M. Kinet, 1996a. Somaclonal variation and improvement in chilling tolerance in rice: changes in chilling-induced electrolyte leakage. Plant Breed. **115** : 268-272

Bertin, P., J. Bouharmont, and J.-M. Kinet, 1997a. Somaclonal variation and improvement of chilling tolerance in rice: changes in chilling-induced chlorophyll fluorescence. Crop Sci. **37**: 1727-1735

Bertin, P., P. Bullens, J. Bouharmont, and J.-M. Kinet, 1998. Somaclonal variation and chilling tolerance improvement in rice: changes in fatty acid composition. Plant Growth Regul. **24**: 31-41

Bertin, P., J.-P. Busogoro, J.-P. Tilquin, J.-M. Kinet, and J. Bouharmont, 1996b. Field evaluation and selection of rice somaclonal variants at different altitudes. Plant Breed. **115**: 183-188

Bertin, P., J.-P. Busogoro, J.-P. Tilquin, J.-M. Kinet, and J. Bouharmont, 1997b. Field performances of rice somaclones at low temperature: effects of callus chilling treatment. Crop Sci. **37**: 1943-1950

Bertin, P., J.-M. Kinet, and J. Bouharmont, 1995. Heritable chilling tolerance improvement in rice through somaclonal variation and cell line selection. Aust. J. Bot. **44**: 91-105

Lutts, S., J.-M. Kinet, and J. Bouharmont, 1995. Changes in plant response to NaCl during development of rice (*Oryza sativa* L.) varieties differing in salinity resistance. J. Exp. Bot. **293**: 1843-1852

Lutts, S., J.-M. Kinet, and J. Bouharmont, 1996a. Effects of salt stress on growth, mineral nutrition and proline accumulation in relation to osmotic adjustment in rice (*Oryza sativa* L.) cultivars differing in salinity resistance. Plant Growth Regul. **19**: 207-218

Lutts, S., J.-M. Kinet, and J. Bouharmont, 1996b. Effects of various salts and of mannitol on ion and proline accumulation in relation to osmotic adjustment in rice (*Oryza sativa* L.) callus cultures. J. Plant Physiol. **149**: 186-195

Lutts, S., J.-M. Kinet, and J. Bouharmont, 1996c. NaCl-induced senescence in leaves of rice (*Oryza sativa* L.) cultivars differing in salinity resistance. Ann. Bot. **78**: 389-398

Lutts, S., J.-M. Kinet, and J. Bouharmont, 1996d. Ethylene production by leaves of rice (*Oryza sativa* L.) in relation to salinity tolerance and exogenous putrescine application. Plant Sci. **116**: 15-25

Lutts, S., J.-M. Kinet, and J. Bouharmont, 1998. NaCl impact on somaclonal variation exhibited by tissue culture-derived fertile plants of rice (*Oryza sativa* L.). J. Plant Physiol. **152**: 92-103

Van Sint Jan, V., C. Costa de Macedo, J.-M. Kinet, and J. Bouharmont, 1997. Selection of Al-resistant plants from a sensitive rice cultivar, using somaclonal variation, *in vitro* and hydroponic cultures. Euphytica. 97: 303-310.

In vitro techniques for the selection of Basmati rice mutants better adapted to saline environments

A.A. Cheema, M.Y. Saleem and M.A. Awan
Nuclear Institute for Agriculture and Biology (NIAB), P.O.Box 128, Faisalabad, Pakistan

Keywords: *Oryza sativa*, salinity tolerance, gamma rays, mutants, *in vitro* screening, induced mutations, somaclonal variation

Summary

Among different doses of gamma rays, maximum callus induction i.e. 71.8% was obtained from 'Basmati Pak' at 15.0 Gy cultured on modified MS medium. 20% callusing response with 7.5, 2.8 and 2.5% differentiation was induced by 20.0 Gy at electrical conductivity (EC) levels of 4, 5 and 6 dS/m of NaCl respectively in MS regeneration medium. A non-irradiated callus, under non-saline environment revealed a differentiation percentage of 48.3% with regeneration percentage of 6.9%, while a non-irradiated saline environment indicated 4% regeneration at 4 dS/m EC of NaCl. In another experiment, mutants were developed from cultivar 'Basmati 370' through the combined use of tissue culture and 20.0 Gy of irradiation dose. Culturing of embryogenic calli onto modified MS regeneration medium supplemented with 25 mg/L ABA slowed down the growth but improved their conversion rate to plants in hormone free MS0 medium. K and NAA promoted the growth of somatic embryos but failed to improve their conversion rate. The combined use of ABA and PGRs (K+NAA) did not trigger plant regeneraiton. In the first somaclonal generation (SC_1), only four plants at 0.2% of NaCl and one at 0.4% of NaCl under gravel culture were screened. The progeny of the screened mutants yielded two salt tolerant mutants with 17% and 33% sterility. The growth value of adapted callus lines at 0.2, 0.25 and 0.3% of NaCl *in vitro* was less than unadapted callus lines, which are under *in vitro* investigation for regeneration potential.

M. Maluszynski and K.J. Kasha (eds.),
Mutations, In Vitro and Molecular Techniques for Environmentally Sustainable Crop Improvement, 161–168.

Introduction

Rice is one of the most important cereal crops and is the staple diet of the population in major rice growing areas of Pakistan. Its annual production is 4.3 million tons; with a foreign exchange return of US $338.5 million (Economic Survey of Pakistan 1996-97).

The twin menace of water logging and salinity are a great threat to irrigated agriculture in Pakistan. About 5.7 million hectares of land have been classified as saline or saline sodic while the area of productive land being damage by salinity increases at a rate of 40,000 ha/year (Imtiaz and Ashraf, 1995), causing an economic loss of US $645 million. Many scientists worldwide believe that available engineering and other options can only minimize the affects of salinity while biologists have emphasized the concept of "adoption and selection" for saline environments.

Stress conditions during plant growth reduce crop yield and impair quality. Following mutation induction, the regenerated plants can be preselected *in vitro* by exposing regenerative cells and tissues to stress conditions, and the preselected genotypes can then be selected for the desired characters in the field. By using somaclonal variation and *in vitro* mutagen induced variability, genotypes tolerant to salt have been obtained in alfalfa (Winicov, 1991; Safarnejad *et al.*, 1996); to drought and soil acidity in sorghum (Duncan *et al.*, 1995), sodium chloride tolerance in tobacco (Nabors *et al.*, 1980), salt and drought tolerance in proso millet (Nabors, 1983) and salt tolerance in sugarcane in Cuba (Maribona, 1994, personal communication). In the present studies, the same approach is being followed by the induction of salt tolerance in Basmati rice.

Materials and methods

Cultivar Basmati Pak

Manually dehusked seeds of rice were irradiated at different doses (0, 5.0, 10.0, 15.0, 20.0 and 25.0 Gy) of gamma rays (^{60}Co). Each set of irradiation treatment and control consisted of 500 seeds along with the control. Both irradiated and non irradiated seeds were sterilized in sodium hypochlorite (NaOCl) for 15 minutes and washed thrice with sterilized distilled water and then cultured onto MS medium (Murashige and Skoog, 1962) modified with 0.15 mg/L benzylaminopurine (BAP) + 30 g/L sucrose. The concentration of the auxin i.e., 2,4-dichlorophenoxyacetic acid (2,4-D) was 1 mg/L throughout the entire studies. The pH was adjusted to 5.8. Later on 10 g/L agar was added to solidify the medium. Fifteen seeds per jar were cultured and all cultures were incubated at 26 ± 2°C in the dark. Observations regarding morphology of compactness of callus and callus percentage were recorded after 4 weeks. Proliferation was done on the same callus induction medium for 4 weeks. Regeneration medium MSR [MS+napthaleneacetic acid (NAA) 0.05 mg/L + BAP 0.05 mg/L + sucrose 30 g/L + agar 10 g/L, pH 5.8] was used. Three salinity levels measured by electrical conductivity (EC), of low, moderate and high (EC4, EC5 and EC6 respectively) were adjusted with NaCl in MSR regeneration medium. 50 ml of medium per jar was autoclaved at 121°C for 15 minutes. Four environments with respect to material and salinity were classified: (1) non-irradiated non-saline (2) non-irradiated saline (3) irradiated non-saline and (4) irradiated saline. Approximately 5 calli per jar (each weighing 50 mg) from both irradiated and non-irradiated material were cultured in each environment under continuous light (2000 Lux) at 26 ± 2°C. Emergence of green shoots (0.1 cm) from callus was taken as an indication of differentiation percentage (D). The number of green plantlets with plant height over 3 cm formed the basis for calculating the percentage of surviving plantlets as regeneration percentage (RG).

Basmati 370

Callogenesis was initiated at room temperature (26 ± 2°C) in the dark from manually dehusked, irradiated (20.0 Gy of gamma rays) and sterilized seeds of 'Basmati 370' on MS medium (Murashige and Skoog, 1962) supplemented with 2 mg/L, 2,4-D, 30 g/L sucrose and 10 g/L agar. The pH of the medium was adjusted to 5.8. After a 30-day incubation period, the calli were subcultured on the same fresh medium for another 30 days. Observations on morphology of the callus were then recorded.

For plant regeneration, the embryogenic calli (50 mg each) were transferred to regeneration medium [MS+2 mg/L kinetin (K) + 1 mg/L napthaleneacetic acid (NAA) + 25 mg/L abscisic acid (ABA) + 30 g/L sucrose + 10 g/L agar]. The pH of the medium was adjusted to 5.8. The cultures were kept in 16 hours light (approx. 2500 - 3000 lux) provided by the fluorescent tubes at a temperature of 26 ± 2°C for 15-20 days. Observations regarding morphology of callus were recorded. Embryogenic calli were then divided into three groups: (1) consisting of 35 somatic embryos per replicate. Regeneration medium MS0 (without plant growth regulators (PGRs) i.e., no K and NAA was used with 30 g/L sucrose and 10 g/L agar, pH 5.8; (2) same as group 1, except regeneration medium (MS + 2 mg/L (K) + 1 mg/L (NAA); (3) same as group 2, except regeneration medium contained 25 mg/L ABA.

The regenerated plants were excised from test tubes and tender roots were washed thrice with distilled water, then transferred to hydroponic. After getting a vigorous root system in the incubator (relative humidity 80%, temperature 28-32°C and fluorescent light 3000 lux) after 15-20 days, the plantlets were then transferred into soil and placed in a growth chamber. At maturity, agronomic characters were recorded. The seeds were bulked to screen for salt tolerant somaclonal variants the next year.

A gravel culture experiment was conducted in a growth chamber to study the effect of salinity i.e., NaCl. Thirty-day old seedlings from the 1st somaclonal generation (SC_1) were chosen for the experiment. The experiment consisted of six treatments of NaCl viz. 0, 0.2, 0.4, 0.6, 0.8 and 1.0% each with three replications. The plants were sown in plastic pots filled with washed gravel containing full strength Hoagland solution. The treatments were created by adding salt solution, starting from 0.1% up to desired concentration. This was done to prevent a sudden shock to plants by salinity. The volume of the solution was maintained by adding Hoagland solution. The plants were harvested upon maturity. Selected plants were evaluated again the next year in the second somaclonal (SC_2) generation in a similar experiment but with five treatments of NaCl: 0.0, 0.4, 0.6, 0.8 and 1.0%. Their salt tolerance was compared at maturity.

Selecting NaCl tolerant or resistant callus lines

Dry dormant, healthy and uniform seeds of Basmati 370 with a moisture content of 14% were incubated on modified MS medium with different levels of 2,4-D (0, 1.5, 2.0, 2.5, 3.0, 3.5 and 4.0 mg/L), 30 g/L sucrose, 10 g/L agar with pH 5.8 for 40-45 days in the dark, at room temperature of 26 ± 2°C. The best performing 2,4-D level was chosen. The callus of the seed embryo was irradiated at 20.0 Gy of gamma rays. After irradiation, embryogenic calli were transferred to modified MS medium with different concentrations (0, 0.2, 0.25 and 0.3%) of NaCl under culture conditions similar to those described for callogenesis for Basmati 370.

Screening of callus lines was conducted through six proliferation phases, each after 20-day intervals with same fresh saline MS medium.

Results and discussion

Basmati Pak

Callusing percentage was increased with increasing doses of irradiation and at 15.0 Gy of gamma rays a maximum callus induction of 71.8% was achieved (Table 1). A similar phenomenon was found by Min *et al.* (1989). Morphology of the callus ranged between embryogenic and non-embryogenic. Differentiation was observed in 48.3% of non-irradiated non-saline and in 4.0% of non-irradiated saline (EC4) environment. In terms of regeneration, the only *in vitro* response was 6.9 and 4.0% in non-irradiated material (Table 1). 10.0 Gy and 15.0 Gy induced 5.0% and 2.0% differentiation respectively in the non-saline environment while at 20.0 Gy, 7.5, 2.8 and 2.5% differentiation was observed in the increasing salinity levels. Different differentiation percentages in non-saline and saline environments may indicate the possibility of getting somaclonal variants at different salinity levels because tissue culture is now being widely used to select salt tolerant plants in various crops.

Table 1. Differentiation and regeneration percentage of non-irradiated and irradiated Basmati Pak under non-saline and saline environments

Radiation dose Gy	Callus induction percentage	Differentiation (%)				Regeneration (%)				Albino plant
		ECo	4	5	6	ECo	4	5	6	
0	46.5	48.3	4.0	-	-	6.9	4.0	-	-	1
5.0	56.5	-	-	-	-	-	-	-	-	-
10.0	58.2	5.0	-	-	-	-	-	-	-	-
15.0	71.8	2.0	-	-	-	-	-	-	-	-
20.0	20.0	-	7.5	2.8	2.5	-	-	-	-	-
25.0	13.4	-	-	-	-	-	-	-	-	-

Basmati 370

The integrated use of tissue culture and irradiation induced 69.5% callusing. Morphology of the callus varied between embryogenic and non-embryogenic callus similar to Min *et al.* (1989). The addition of ABA in combination with plant growth regulators was added to improve the conversion rate but it showed an inhibitory action (Von Arnold and Hakman, 1988). Fifty percent differentiation and 30% regeneration (12 plants) in addition to four albino plants was found from Group (1) (Table 2). This indicates that ABA alone or in combination

Table 2. Conversion rate of somatic embryos from callus of Basmati 370 with different plant growth regulator

Group	Medium	Embryo-genic calli (No.)	Green shoot (No.)	Differentiation (%)	Albino plants (No.)	Green plants (No.)	Regeneration (%)
Group(1)	MS0						
1st subculture	"	40	20	50	4	12	30
2nd "	"	28	-	-	-	-	-
3rd "	"	30	-	-	-	-	-
Group (2)	MS+PGRs						
1st subculture	"	50	20	40	-	-	-
2nd "	"	43	-	-	-	-	-
3rd "	"	20	-	-	-	-	-
Group (3)	MS+PGRs+ABA						
1st subculture	"	50	-	-	-	-	-
2nd "	"	31	-	-	-	-	-
3rd "	"	5	-	-	-	-	-

with PGRs inhibits regeneration but apparently enhances conversion rate in hormone free MS0 medium (Isabelle *et al.*, 1993). In Group 2, K and NAA promoted growth of somatic embryos with 40% differentiation but did not improve the conversion rate to plants. Growth regulators appear to have different effects on somatic embryos or embryogenic tissue, depending on the ontogenic stage of the target tissue (Fujimura and Komamine, 1980). In Group 3, when applied in combination with ABA, the PGRs (K+NAA) did not trigger plant recovery even after three consecutive subcultures on the same regeneration medium (Table 2). The maturity

Table 3. Agronomic traits of SC_1 generation of Basmati-370 grown under different salinity levels

NaCl level (%)	Plant No.	Plant height (cm)	Days to flowering	Tillers (No.)	Panicle length (cm)	Spikelets/ panicle (No.)	Sterility (%)	Yield/ plant (g)
0.0	1	110	115	8	23.5	80	49	1.50
	2	90	118	8	20.0	80	38	1.93
	3	93	114	9	22.5	83	37	1.45
	4	102	113	10	22.0	84	30	1.41
0.2	1	92	112	7	24.0	83	58	1.60
	2	100	111	7	18.0	76	77	1.52
	3	93	114	8	24.5	83	83	1.75
	4	93	118	6	23.0	63	63	1.04
0.4	1	98	114	9	23.0	94	30	0.64
	2							
	3							
	4							
0.6	Complete sterility							
0.8	Lethal							
1.0	Lethal							

period of all somaclonal plants was similar (Table 3). Considerable variation was observed in plant height i.e., 52-94 cm. Variation in the size and color of grain was also found similarly to Shen *et al.* (1995). Complete sterility was observed at 0.6% of NaCl while at 0.8 and 1.0% of NaCl complete lethality was observed. Screening of four plants in 0.2% and one in 0.4% of NaCl gravel culture was done. As pointed out by Stavarek and Rains (1984) plants which regenerated under salinity stress often had variable morphology and were often sterile. There was no remarkable difference between yield and yield components. At 0.4% of NaCl level, the yield of selected plants decreased to 0.64 g because of increased osmotic level (NaCl). Despite the fact that the other three plants died, the sterility was only 30%. So this plant should be evaluated for genetic stability of traits. Ahloowalia (1997) suggested that it is necessary to study the genetic stability of somaclonal variation in at least two subsequent seed generations. The seed of each plant was collected to evaluate the next generation for salt tolerance. In SC_2 generation two somaclonal mutants of SC_1 previously screened at 0.4% of NaCl exhibited 17 and 33% sterility respectively, having yield 0.9 and 3.1 g per plant (Table 4). The comparison

Table 4. Agronomic traits of SC_2 generation of Basmati 370 grown under different salinity levels

NaCl level	Previous level of NaCl	Plant No.	Plant height (cm)	Days to flowering	Tillers (No.)	Panicle length (cm)	Spikelets per panicle (No.)	Sterility (%)	Yield per plant (g)
0.0	0.2	1	112	110	6	21	85	59	1.4
		2	111	114	5	25	68	60	2.3
		3	114	115	4	24	73	64	1.8
0.4	"	1	118	118	3	23	80		
		2	114	114	1	14	54	100	-
		3	-	-	-	-	-	-	-
0.6	"	1	113	116	4	21	80	98	-
		2	114	108	4	22	79	97	-
		3	-	-	-	-	-	-	-
0.8	"	1	115	113	2	25	111	88	0.15
		2	-	-	-	-	-	-	-
		3	-	-	-	-	-	-	-
1.0% Lethal treatment									
0.0	0.4	1	113	112	4	21	82	14	7.9
		2	115	112	3	24	91	85	0.8
		3							
0.4	"	1	116	115	6	25	86	17	0.9
		2	114	114	5	20	76	33	3.1
		3	-	-	-	-	-	-	-
0.6	"	1	113	113	6	17	74	76	0.9
		2	115	111	2	24	78	99	-
		3	-	-	-	-	-	-	-
0.8	"	1	115	112	4	18	59	100	-
		2	116	110	3	19	64	-	-
		3	-	-	-	-	-	-	-
1.0	Lethal treatment								

of sterility and yield of the parent plant selected at 0.4% NaCl and of two of its' selected plants from SC_2 generation confirmed the genetic instability of the traits and promises to select salt tolerant plants with low sterility and high yield in the succeeding generation. At 0.6% of NaCl level and beyond, there was a drastic increase in sterility percentage indicating the highly sensitive nature of Basmati type varieties to salt.

Screening of salt tolerant/resistant callus lines

Maximum callus induction was observed with auxin; 2,4-D at the rate of 2 mg/L in MS medium which was taken as standard for experimentation (Raina *et al.*, 1987). Growth value of unadapted callus lines was greater than adapted callus lines in a saline environment. At 0.2, 0.25 and 0.3% of NaCl out of 235, 180 and 30 unadapted cell lines, 165, 105 and 30 adapted lines were selected, respectively. A similar level was reported by Basu *et al.* (1997). These adapted callus lines have been transferred to a regeneration medium.

References

Ahloowalia, B.S., 1995. *In vitro* mutagenesis for the improvement of vegetatively propagated plants. In: Induced Mutations and Molecular Techniques for Crop Improvement. IAEA, Vienna. pp.531-541

Basu, S., G. Gangopadhyay, B.B. Makherjee, and S. Gupta, 1997. Plant regeneration of salt adapted callus of *indica* rice (var. Basmati 370) in saline conditions. Plant Cell Tiss. Org. Cult. **50**: 153-159

Duncan, R.R., R.M. Waskom, and M.W. Nabors, 1995. *In vitro* screening and field evaluation of tissue-culture-regenerated sorghum (*Sorghum bicolor* (L.) Moench) for soil stress tolerance. Euphytica. **85**: 373-380

Fujimura, T. and A. Komamine, 1980. Mode of action of 2,4-D and Zeatin on somatic embryogenesis in carrot cell suspension culture. Z. Pflanzenphysiol. **99**: 1-8

Imtiaz, A. and M. Ashraf, 1995. Saline agriculture for enhanced food production. Progressive Farming. **15**: 41-45

Isabelle, G.T., M.C. Mauro, L. Sossuntzoo, E. Migniac, and A. Detorie, 1993. Arrest of somatic embryo development in grapewine: histological characterization and the effect of ABA, BAP and zeatin in stimulating plant development. Plant Cell Tiss. Org. Cult. **3**: 91-93

Min, S., X. Qi, Z. Xiong, and Ch. Zhao, 1989. Effects of gamma-radiation treatment in somatic cell culture of indica rice Basmati 370 selection. In: Proc. of the 6th Internatl. Congr. of SABRAO. pp.793-796

Murashige, T. and F. Skoog, 1962. A revised medium for rapid growth and bioassays with tobacco tissue cultures. Physiol. Plant. **15**: 473-497

Nabors, M.W., 1983. Increasing the salt and drought tolerance of crop plants. In: Current Topics in Plant Biochemistry and Physiology. Randolph, L.F. (Ed.) Univ. Missouri Press, Columbia. pp.165-184

Nabors, M.W., S.E. Gibbs, C.S. Bernstein, and M.E. Meis , 1980. NaCl tolerant tobacco plants from cultured cells. Z. Pflanzenphysiol. **97**: 13-17

Raina, S.K., P. Sathish, and K.S. Sarma, 1987. Plant regeneration from *in vitro* cultures of anthers and mature seeds of rice (*Oryza sativa* L.) cv. Basmati-370. Plant Cell Rep. **6**: 43-45

Stavarek and Rains, D.W., 1984. The development of tolerance to mineral stress. HortSci. **19**: 377-382

Safarnejad, A., H.A. Collin, K.D. Bruce, and T. McNeilly, 1996. Characterization of alfalfa (*Medicago sativa*) following *in vitro* selection for salt tolerance. Euphytica. **92**: 55-61

Shen, Y., Q. Cai, M. Gao, and Z. Liang, 1995. Isolation and genetic characterization of somaclonal mutants with large-sized grain in rice. Cereal Res. Commun. **23**(3): 235-241

Von Arnold, S. and L. Hakman, 1988. Regulation of somatic embryo development in *Picea abies* by abscisic acid (AB). J. Plant Physiol. **132**: 164-169

Winicov, I., 1991. Characterization of salt tolerant alfalfa (*Medicago sativa* L.) plants regenerated from salt tolerant cell lines. Plant Cell Rep. **10**: 561-564.

Rice germplasm enhancement by induced mutations in Chile

J.R. Alvarado Aguila and R.P. Madariaga Burrows
Centro Regional de Investigación Quilamapu, Instituto de Investigaciones Agropecuarias (INIA) Casilla 426, Chillán, Chile

Keywords: cold tolerance, earliness, disease resistance, *Oryza sativa*, grain quality, gamma rays, mutants

Summary

The main goals of the rice breeding program in Chile are: cold tolerance, earliness, long and translucent grain and disease resistance. It has been very difficult to combine in one genotype the long and translucent grain with earliness. Therefore, six varieties or the selected Chilean lines: 'Diamante-INIA', 'Cinia 606', 'Buli-INIA', 'Cinia 609', 'Quella-INIA' and 'Quila 31808' were irradiated with 200 Gy of gamma rays at the Chilean Atomic Energy Agency. The main objective was to induce greater variability in order to select for earliness and to breed for resistance against sheath and stem rot. The results have shown that some mutant lines of varieties Diamante-INIA, Buli-INIA and lines Cinia 606 and Cinia 609 are at least 6 days earlier than the parent varieties or lines. During 1997/98 under field conditions, plants were inoculated with *Rhizoctonia oryzae sativae.* Great variability was found in Cinia 609, in relation to the length of lesion after the inoculation and seven M_3 families had lesions less than 1 cm in length.

M. Maluszynski and K.J. Kasha (eds.),
Mutations, In Vitro and Molecular Techniques for Environmentally Sustainable Crop Improvement, 169–173.

Introduction

With around 30,000 hectares grown annually, rice is one of the traditional crops in Chile. It is cultivated under flooded conditions from sowing to maturity. The Chilean rice area is located between 34°10' S.L. and 36°34' S.L. in the irrigated central valley. The research program is located in the south border at Quilamapu Regional Research Center, INIA, Chillán. The weather of the rice area is temperate with a Mediterranean climate (Alvarado, 1991). The main constraints to Chilean rice are: cold temperature, soil leveling, soil fertility, weeds, water management and the severity of sheath and stem rot caused by *Rhizoctonia oryzae sativae* and *Sclerotium hydrophilum*. Thus, the principal goals of the rice breeding program are: cold tolerance, earliness, long and translucent grain, grain quality and disease resistance (Alvarado, 1997).

Rice plant breeding and agronomy programs still have many problems without answers. The long grain varieties are late and they lack significant resistant to stem and sheath rot. Rice yield potential has not changed for many years and for the best production, the farmers' genotypes usually have similar yield potential.

Since rice breeding started in Chile, the major problem has been the lack of variability in the germplasm for cold tolerance, earliness, grain quality, and disease resistance. It is known that the success of induced mutations is based on using well adapted varieties needing only one or two genetic improvements of agronomic characters (Rutger, 1991) and that earliness is a factor for which induced mutations have been very successful (Rutger, 1983; Anon, 1990; Maluszynski *et al.*, 1995). With conventional breeding, it has been very difficult to combine in one genotype, long and translucent grain with earliness. Therefore, induced mutations appeared to be a good method for producing this combination. The principal goals of the mutation project was to induce greater variability, select earliness and initiate breeding for resistance to sheath and stem rot.

Material and methods

The rice varieties and advanced lines used were: 'Diamante-INIA' and 'Cinia 606' (with long and translucent grain and a length-width ratio less than 3); 'Buli-INIA' and 'Cinia 609' (with long and translucent grain and a length-width ratio greater than 3); and 'Quella-INIA' and 'Quila 31808' (with short grain and a length-width ratio less than 2). Seeds of all of them were irradiated with 200 Gy gamma rays at the Chilean Atomic Energy Agency.

During 1995, the genotypes (M_1, varieties and lines) were sown presoaked in to water on November. Two plots, 1 x 7 m, per genotype were sown with irradiated seeds and one check plot with the respective variety or line with untreated seeds. The seeding rate was 300 kg/ha. Rice was fertilized with N-P-K at the rates of 90, 45 and 36 kg/ha of N, P_2O_5 and K_2O, respectively. During 1996, mutated material, varieties and lines were sown presoaked in to water in October. Two plots, 1x14 m, per genotype were sown with M_2 seeds and one plot with the respective parental material. The M_2 mutants were transplanted in November. Rice was fertilized with N-P-K and the rates used were 160, 80 and 60 kg/ha of N, P_2O_5 and K_2O, respectively.

In 1997/98 season, the seeds collected from M_2 generation were presoaked in September and the M_3 plants were transplanted 18 days later. The weeds were controlled with Molinate and Sirius herbicides.

In the case of resistance to sheath and stem rot, mutated material and parental plants were inoculated with *Rhizoctonia oryzae sativae.* The fungus utilized was the isolate Ros 97.2, obtained from sheaths of rice with symptoms. The isolate was maintained at 5°C. Before its utilization, it was transferred to PDAA medium and then incubated in darkness for four days at 24 °C. Two groups of plants were inoculated during the 1996/97 season, one under field conditions and other under greenhouse conditions. Under controlled conditions the plants were compared with those from parental, unirradiated seed. The artificial inoculation was done at the end of effective tillering stage. The evaluation was conducted 7, 14 an 21 days after the inoculation. Fifty mutant lines corresponding to selected plants under field conditions, 12 selected mutant lines from controlled conditions and a group of 19 elite lines and checks were again evaluated during the 1997/98 season. Two inoculations were made during 1997/98, one on the14th and the second on the 23rd of January, using different tillers of the same plants. Each inoculated tiller was tagged for identification. Each tiller received 0.4 ml of the inoculum solution. At maturity each tiller was evaluated, by weighing the grains produced and measuring the length of the lesions.

Results

Selection for earliness

During two seasons, 1996/97 and 1997/98, individual plants were selected for earliness and other agronomic traits (Table 1). During the 1997/98 crop season, the climatic condition were unfavorable for rice production, due to low temperatures during the full growing season. The average temperature from October to March was the lowest in the last twenty years, reaching 16.8°C in the southern part of the rice area. However, it was possible to select M_3 plants with earliness and good spikelet fertility. Some early lines were lost due to infection by stem rot, particularly the mutant lines from Quella-INIA. Normally the difference in the period from sowing to flowering between Oro, the most popular old variety and considered by farmers to be an early variety, and Diamante-INIA is around 7 days. Buli-INIA, Cinia 606 and Cinia 609 have a similar flowering period to Diamante. This difference has been enough to escape, in some seasons, the low temperatures during the flowering stage.

Table 1. Number of rice plants selected for earliness from an irradiated population in two successive crop seasons and generations

Variety or line	No. of selected plants	
	1996/97 (M_2)	1997/98 (M_3)
Diamante-INIA	158	100
Buli-INIA	70	34
Quella-INIA	34	6
Quila 31808	--	--
Cinia 606	66	35
Cinia 609	48	28
Total	376	203

In the 1997/98 season Diamante-INIA, needed 90 days from transplanting to flowering while mutant lines selected from this parent ranged from 83 to 93 days. At least 9 mutant lines flowered 6 to 7 days earlier than Diamante-INIA. The parental Buli-INIA took 100 days from transplanting to flowering, while its mutant lines varied from 88 to 100 days. Twenty-two mutant lines were 7 days earlier than their parent variety. The breeding line Cinia 606 reached flowering in 92 days whereas the mutant lines varied from 83 to 91 days. Five lines were 8 days earlier than the parental line.

Only 3 mutant lines from Cinia 609 were 6 days earlier than parent. The 1997/98 crop season was characterized by low temperature during the flowering period and 295 mutants were selected for a high spikelet fertility (Table 2). These results show that the selection of irradiated rice having long and translucent grains is a good method to increase rice variability for earliness along with long and translucent grains.

Table 2. Number of mutants selected for high spikelet fertility from mutated generations in 1997/98 growing season

Genotypes	No. of mutants
Diamante-INIA	88
Buli-INIA	87
Quella-INIA	4
Cinia 609	54
Cinia 606	31
Quila 31808	31
Total	295

Selection of mutants resistant to Rhizoctonia oryzae sativae

In the 1996/97 growing season, irradiated and not irradiated progenies of rice lines and varieties were planted and inoculated with *Rhizoctonia oryzae sativae* under greenhouse conditions. The artificial inoculation was performed when the rice plants were at the end of the effective tillering period. The results showed that one mutant plant of the Cinia 609 line did not have lesions following inoculation. In contrast, the other inoculated plants had lesions of different lengths. The average length of lesions in parental varieties ranged from 1.60 cm (Cinia 606) to 3.25 cm (Quella-INIA).

In the next season (1997/98), the selected lines were also studied. In this year, the inoculation was done at earlier stages of rice development. The behavior of mutated lines was different in the two seasons. For example, lines from Cinia 609 performed with higher resistance in 1996/97 and lower resistance in 1997/98 and lines from Cinia 606 had an opposite behavior. By analysis of each individual plant over two years, it was found that one mutated line of Diamante-INIA had a lesion length of less than 1 cm.

In a second experiment, lines of Cinia 609 that had originated from selected plants inoculated under field conditions in 1996/97 were studied. In the mutated generation, the length of lesions showed great variability, having plants with lesions smaller than 1 cm to plants with lesions larger than 7 cm (Fig. 1). Seven M_3 families had plants with lesions less than 1 cm. These families will be studied in the next growing season. In relation to the elite lines studied, the lesions ranged from 1.4 cm to 3.7 cm. Cinia 807 showed the lowest value and Cinia 244 presented the highest value relative to length of the lesions caused by the inoculation.

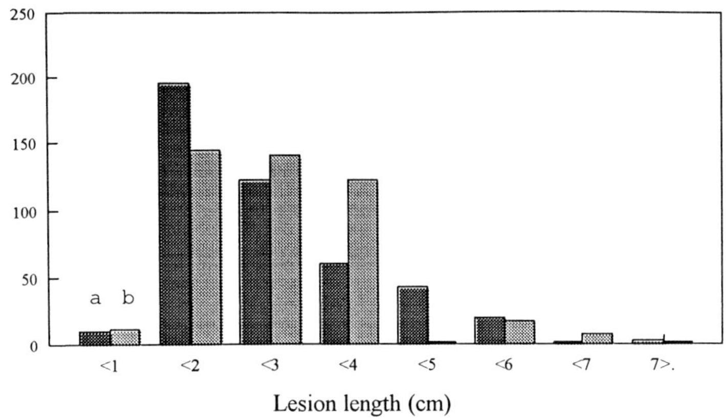

Figure 1. Distribution of mutants from Cinia 609 according to the length of the lesion in 2 dates of inoculation in 1997/98. a - January 14; b - January 23

References

Anon. 1990. Centre Francais du Riz. Ses activites de recherche et de development. Mas du Sonnailler. Arles. pp.36

Alvarado, J.R., 1991. Mejoramiento de arroz en Chile. In: Mejoramiento de Arroz. Puignau, J. (Ed.) IICA, Montevideo. pp.51-61

Alvarado, J.R., 1997. Mejoramento de arroz en Chile y uso de seleccion recurrente. In: Selection Recurrente en Arroz. Guimaraes, E. (Ed.) CIAT, Cali. pp.117-123

Maluszynski, M., L. van Zanten, A. Ashri, H. Brunner, B.S. Ahloowalia, F.J. Zapata, and E. Weck, 1995. Mutation techniques in plant breeding. In: Induced Mutations and Molecular Techniques for Crop Improvement. IAEA, Vienna. pp.489-504

Rutger, J.N., 1983. Applications of induced and spontaneous mutation in rice breeding and genetics. Adv. Agron. **36**: 383-413

Rutger, J.N., 1991. Mutation breeding of rice in California and the United States of America. In: Plant Mutation Breeding for Crop Improvement. Vol. 1. IAEA, Vienna. pp.155-165.

Induced mutations as a method of obtaining iron toxicity resistant and high quality rice cultivars

R.E. Bacha, S. Yokoyama and T. Ishiy
Epagri – Estação Experimental de Itajaí, Itajaí, SC, Brazil

Keywords: Oryza sativa, soil acidity, grain quality, cooking quality, gamma rays, mutants

Summary

Iron toxicity, a nutritional disorder of rice plants typical of acid soils rich in iron oxides, has been detected as a yield reducing factor in several countries of Asia, Africa and South America. Iron toxicity in Santa Catarina, Brazil, has been responsible for yield reductions up to 80%. Soil amendments and cultural practices are not recommended to correct the problem. Tolerant or medium resistant cultivars are the best solution to be used. An attempt to obtain germplasm resistant to iron toxicity and good quality rice has been made using induced mutations. Gamma rays (120-240 Gy) were used as a physical mutagen to improve 'EMPASC 105', an iron susceptible and poor grain quality rice cultivar. Plants from M_3 up to M_6 were field screened for tolerance to iron toxicity. Agronomic characteristics, milling yield and cooking quality were also evaluated. One mutant line, SCM 7-12-4, from EMPASC 105 was selected for its medium resistance to iron toxicity, lodging resistance, high milling yield and good cooking qualities. The mutant line was evaluated in the first year, of the three-year regional trials, under six different soil and climatic conditions.

M. Maluszynski and K.J. Kasha (eds.),
Mutations, In Vitro and Molecular Techniques for Environmentally Sustainable Crop Improvement, 175–182.

Introduction

In the early 70's, semidwarf rice varieties were introduced in Brazil from IRRI (International Rice Research Institute) and CIAT (Centro Internacional de Agricultura Tropical). Iron toxicity, a nutritional disorder of rice plants characteristic to acid soils (Ponnamperuma and Castro, 1972), was then observed with the newly introduced rice germplasm. Depending on soil type and varietal susceptibility, the final grain yield may be reduced up to 80% due to iron toxicity. Chemical methods and other cultural practices may not entirely solve the problem, nor be economically feasible (Chandler, 1979). The use of tolerant or resistant cultivars, therefore, is the most effective and inexpensive way to overcome the problem.

Breeding and induced mutations may provide sources of germplasm for tolerance to iron toxicity and good quality rice (Ishiy, 1991). A screening program for iron toxicity has been performed since 1984 at Itajaí Experiment Station-Epagri, located in Southern Brazil (Bacha, 1991). Between 1984 and 1997, 10,900 genetic lines have been screened for iron toxicity. Some high yielding cultivars have been released, even though they present agronomic and grain quality problems. Rice cultivar 'EMPASC 105' is a high yielding rice germplasm tolerant to *Pyricularia grisea* (blast) and is non-lodging. However, it has poor grain quality (stickiness after cooking) and susceptibility to iron toxicity. Resistance to lodging is very important to rice growers of the State of Santa Catarina, since they use the water seeding system. However, stickiness is not generally acceptable in the Brazilian market, except to the Asiatic communities. The objectives of this research are to obtain rice genotypes with good grain quality and resistance to iron toxicity, using gamma radiation induced mutations.

Materials and methods

Seeds of the cultivar EMPASC 105 were irradiated with gamma rays at 120-240 Gy in 1994. M_1 single plants were hand-transplanted and three panicles per plant were harvested. M_2 single plants were also transplanted and *Pyricularia* resistant plants were selected. M_3 single plants were the first evaluated for iron toxicity tolerance. From the M_4 populations, 150 mutant lines were selected and sown as M_5 in the iron-screening nursery in 1995. Symptoms of indirect toxicity, expressed by yellow or orange colors of the leaves (Howeler, 1973), develop in the susceptible and medium resistant plants due to accumulation of iron oxides on rice root surfaces (Bacha and Hossner, 1977). Evaluations were performed from 30 to 90 days after continuous flooding. According to the discoloration intensity, or death of the leaves, the lines were rated from 0 to 9 with 0 - 3 being resistant (R), 4 - 5 being medium-resistant (MR), and 6 - 9 being susceptible (S), according to IRRI (1976). Resistant plants were selected at the nursery and seeds were harvested to form the M_6 new entry lines.

Grains of the new lines were analyzed for grain quality, namely amylose content, gelatinization temperature, milling yield and cooking quality. Field characteristics were determined as yield components, life cycle, panicle exertion, and blast tolerance (EMBRAPA, 1977).

After being screened in 1995-96, for iron toxicity and other characteristics, 24 mutant lines (M_6) were selected for planting in 1996-97. From these, 14 lines were discarded for several agronomic characters, such as plant architecture, kernel malformation, poor panicle exertion and cooking quality, and 10 lines were selected and evaluated for grain yield, iron toxicity, tillering, milling quality, cooking quality and blast resistance.

In 1997-98, one mutant line, SCM 7-12-4, was included in the regional trials in six locations of Santa Catarina. These locations were selected as being representative of the soil and climate of different areas where rice is cultivated. Five areas represent mineral soils and one area represents organic soils. Fifteen lines per trial were water-seeded in 60 m^2 plots. Each trial was conducted according to the recommendations in the rice technical bulletin for the State. As a normal procedure, each entry must stay at least three years in the regional trials before it is released as a new cultivar. However, it can be eliminated from the trials after the second year if it does not meet the minimum yield requirement of 7.0 t/ha in one location.

Results and discussion

The results (Table 1) show that most mutant lines (SCM) had either susceptible or medium resistance reactions to iron toxicity. Compared to the parent cultivar EMPASC 105, only three lines, SCM 7-15-2, SCM 8-45-1 and SCM 7-12-4, indicated some improvement. This was a preliminary result because the intensity of symptoms of iron toxicity may vary according to climate conditions in different planting seasons (Bacha, 1989). However, all the other cultivars, were consistent with their expected resistance/tolerance to iron toxicity.

Blast infection is another trait that is influenced greatly by climate and some cultural practices. The data suggest that a very mild infection occurred during the 1995/96 season because even the susceptible variety, EMPASC 105, presented a score of resistance to blast.

The number of tillers was high for all materials and all mutant lines were early maturing compared with the check EMPASC 105, which is medium maturing. Duration of the life cycle is an easily mutated character (Syakudo and Yamagata, 1960). However, the shortening in cycle had a detrimental effect on grain yield, as the grain yield of the early mutants (Table 1) was always lower than that of the parent cultivar.

Gelatinization temperature and amylose content are the two main values that measure grain quality (Webb, 1994). According to the data presented in Table 1, life cycle and amylose content were the two most evidently mutated characters. Gel temperature did not mutate as much as amylose content. Genetic mutations of EMPASC 105 have generated, in this case, genotypes that combine good gel temperature and amylose content. Intermediate values for both amylose and gel temperature are the desired values; however, for the internal market, rice with intermediate amylose and high gel temperature is acceptable. Both low amylose and low gel temperature will result in stickiness after cooking, which is not marketable in Brazil.

The three lines SCM 7-15-2, SCM 7-19-1 and SCM 7-12-4 hold interest because of their final cooking characteristics. They presented the same cooking quality as the three commercial cultivars 'EPAGRI 106', 'EPAGRI 107' and 'EPAGRI 109'. The mutant line SCM 7-15-2 was discarded from future evaluations due to its tendency to lodge, and SCM 7-19-1 showed susceptibility to iron toxicity. All lines presented high milling yield. Although the line SCM 7-12-4 yielded less than EMPASC 105, it was selected for its high whole grain yield (68.80%), intermediate amylose content (I), high gelatinization temperature (H), early maturity (120 days), resistance to lodging and tolerant reaction to iron toxicity. The line SCM 7-12-4 meets all field and lab requirements for a line to be evaluated in the regional trials.

The results of the regional trials conducted in six locations during 1997-98 are presented in Tables 2 and 3. Except for evaluations of iron toxicity and lodging, these data represent the average observations of the five locations. One trial was completely eliminated from the evaluations based on its poor initial stand. One location may determine if the

Table 1. Iron toxicity, grain quality and other agronomic characters of ten rice mutants selected in the 1995-96 season, compared to four rice cultivars

Genotype	Iron toxicity[1]	Blast tolerance[2]	Tillers/plant (No.)	Days to maturity[3]	Amylose content[4] (%)	Gel temp.[5]	Milling (%)[6] W	Milling (%)[6] B	Yield (t/ha)
01-SCM 7-4-4	6	1	18.6	120	I	L	66.04	4.95	6.20
02-SCM 7-6-1	6	1	13.5	120	L	L	64.03	5.83	5.94
03-SCM 7-15-2	5	1	14.4	120	I	H	60.97	8.20	5.87
04-SCM 7-13-1	6	1	12.3	120	L	L	64.64	4.84	5.33
05-SCM 7-31-1	6	2	14.1	120	L	L	64.34	4.71	5.00
06-SCM 7-19-1	7	1	13.0	120	I	H	61.45	7.75	5.52
07-SCM 7-13-2	7	3	13.3	125	L	L	63.35	6.04	5.42
08-SCM 8-52-1	6	1	14.8	128	I	L	-	-	5.04
09-SCM 8-45-1	5	1	16.6	125	I	L	64.63	3.95	3.64
10-SCM 7-12-4	5	1	14.7	120	I	H	68.80	4.39	5.90
11-EMPASC 105	6	1	14.7	135	L	L	63.35	8.53	7.22
12-EPAGRI 106	4	2	13.2	120	I	H	64.03	6.31	6.40
13-EPAGRI 107	3	1	13.7	128	I	H	63.30	8.70	6.50
14-EPAGRI 109	3	1	19.1	140	I	H	67.32	4.00	8.30

[1/] Iron toxicity - (0-3) resistant, (4-5) medium-resistant, (6-9) susceptible; [2/] Blast resistance - (0-3) resistant, (4-6) medium-resistant, (7-9) susceptible; [3/] Days to maturity - From germination to maturity; [4/] Amylose content - L = low, I - intermediate; [5/] Gelatinization temperature – L = low, H = high; [6/] Milling - W = whole grain, B = broken.

Table 2. Agronomic evaluations and grain quality characteristics of genetic lines in regional trials in Santa Catarina, Brazil (1997-98)

Germplasm	Plant height (cm)	Days to maturity[1]	Grains/ panicle[2]	1000 grains weight (g)	Iron toxicity	Lodging[3]	Amylose[4] (%)	Gel[5] temperature	MilingYield[6]	
									W	B
SC 149 (2)[1]	83	104	54	28	S	S	H (28)	L (7,0)	63.1	5.5
SC 152 (2)	81	120	96	28	S	MR	I (26)	H (3,0)	65.2	3.8
SC 154 (1)	102	120	89	25	R	R	I (27)	H (2,0)	54.7	5.3
SC 153 (1)	94	120	99	27	S	S	I (27)	L (7,0)	64.5	6.9
SC 155 (1)	88	118	81	27	MR	S	I (26)	H (2,0)	67.9	2.6
SC 157 (1)	98	107	92	30	MR	S	I (27)	L (7,0)	59.8	6.5
SC 159 (1)	96	107	86	28	S	MR	L (21)	L (6,0)	62.4	4.0
SCM7-12-4	96	120	98	27	MR	R	I (27)	H (2,0)	65.5	2.0
EPAGRI 106 (C)	88	115	110	26	MR	MR	H (29)	H (2,0)	63.2	8.7
SC 158	94	130	96	30	S	S	I (27)	L (6,0)	56.8	8.7
EPAGRI 107 (C)	98	125	113	23	R	R	I (27)	H (2,0)	54.3	16.7
SC 151 (2)	88	135	97	33	S	MR	I (27)	H (2,0)	68.6	3.9
SC 156 (1)	102	142	77	31	S	R	I (27)	I (4,0)	58.9	9,6
SC 160 (1)	98	145	101	28	MR	R	H (30)	H (2,0)	63.3	5.0
EPAGRI 108 (C)	98	145	92	29	R	R	I (27)	H (2,0)	65.1	6.1

1/ Number of years in the trials; 2/ Days from emergence to maturity; 3/ Number of full grains per panicle; 4/ S = susceptible, MR = medium-resistant, R = resistant; 5/ H = high, I = intermediate, L = low; 6/ W = whole grain, B = broken

Table 3. Grain yield of 15 rice germplasm of irrigated rice in five locations of Santa Catarina State, Brazil (1997-98)

Germplasm	Location[2]					Mean (t/ha)
	1	2	3	4	5	
Early maturing						
1 SC 149 (3)[1]	7.3	7.1	0.0	4.4	5.4	6.1
2 SC 152 (3)	6.9	8.1	5.9	5.3	5.8	6.4
3 SC 154 (2)	7.5	7.9	5.1	5.5	5.8	6.4
4 SC 153 (2)	8.0	7.6	0.0	6.4	6.5	7.1
5 SC 155 (2)	7.7	8.6	4.6	6.2	5.9	6.6
6 SC 157 (1)	6.7	6.2	0.0	5.3	4.2	5.6
7 SC 159 (1)	6.2	7.6	0.0	5.0	5.5	6.1
8 SCM7-12-4(1)	6.2	7.7	5.2	5.7	4.9	5.9
Mean per location	7.1	7.6	5.2	5.5	5.5	6.2
9 EPAGRI 106 (Check)	7.0	7.1	0.0	4.6	5.9	6.1
Medium						
10 SC 158 (1)	8.8	8.6	5.7	6.1	6.5	7.1
11 EPAGRI 107 (Check)	7.7	7.4	4.1	4.8	5.6	5.9
Late						
12 SC 151 (3)	7.7	7.6	5.3	4.6	7.1	6.5
13 SC 156 (2)	7.1	6.8	0.0	0.0	5.2	6.4
14 SC 160 (1)	6.9	7.6	5.3	0.0	5.2	5.0
Mean per location	7.2	7.3	5.3	4.6	5.8	6.0
15 EPAGRI 108 (Check)	7.9	9.1	6.3	6.5	6.5	7.3

[1/] Number of years in the trial; [2/] 1-Itajaí; 2-Pouso Redondo; 3-Joinville; 4-Tubarão; 5-Turvo

genotype is susceptible or medium resistant to lodging. To be classified as resistant, the germplasm has to be resistant in all locations. Under normal conditions, lodging is an eliminatory determination in every step before being released as a new cultivar. During abnormal years, such as the 1997-98 season with the "El Niño" phenomenon, some germplasm that could have been classified as resistant to lodging, may have now been seen as susceptible. High rainfall, strong winds and cloudy days were frequent during this season.

The three cultivars representing the checks for early, medium and late maturing germplasm were EPAGRI 106, EPAGRI 107 and 'EPAGRI 108' respectively. The 15 genotypes (Table 2) are grouped according to their life cycle duration and all are classified as long grain.

The mutant line SCM 7-12-4 presents some advantages over the check, EPAGRI 106, with respect to resistance to lodging, milling yield and cooking qualities. EPAGRI 106 tends to remain hard after cooking as a function of its high amylose content. The intermediate value of amylose in the mutant line gives the rice softness after cooking. The general grain yield was lower than in normal years, and this is attributed to adverse climatic conditions. Yields over 9 t/ha are not uncommon among farmers in the State of Santa Catarina. From six initial trials, results were obtained only in five locations due to the total loss that occurred in one location. Table 3 shows the grain yield obtained in the regional trials during the 1997-98 season. Locations 1 and 2 are the sites where grain yields are normally high; location 4 holds the trial on organic soils. Zero yields were due to damage to the plots caused by intense rain or flash floods.

The average yield of the early maturing germplasm was similar to those obtained with the medium and late line. Higher yields are achieved, normally, with late germplasm. The best high yielding cultivar, EPAGRI 108, did not exhibit its yield potential of over 12 t/ha; on the contrary, the mean yield of 6.0 ton/ha obtained in the trials equaled that of the mutant line SCM 7-12-4. Conclusive results on the performance of SCM 7-12-4 will await its evaluation during the next two years of tests on regional trials.

References

Bacha, R.E., 1989. Avaliação de linhagens e cultivares de arroz irrigado para condições adversas de solo: toxidez por ferro. In: Reuniào da Cultura do Arroz Irrigado, 18, Porto Alegre. Anais. IRGA, Porto Alegre. p. 247-252

Bacha, R.E., 1991. Avaliaçào de linhagens e cultivares de arroz irrigado para condiçòes adversas de solo: toxidez por ferro. In: Reuniào da Cultura do Arroz Irrigado, 19, Balneário Camboriú. Anais. EMPASC, Florianópolis. p. 156-159

Bacha, R.E., and L.R. Hossner, 1977. Characteristics of coatings formed on rice roots as affected by iron and manganese additions. Soil Sci. Soc. of America J. **41**:931-935

Chandler, R.F., 1979. Promising rice research. In: Chandler, R.F. (Ed.) Rice in the Tropics: a Guide to the Development of National Programs. West-View Press, Boulder, Colorado. p. 145-164

Empresa Brasileira de Pesquisa Agropecuária-EMBRAPA, 1977. Manual de Métodos de Pesquisa em Arroz. EMBRAPA/CNPAF, Goiânia, Go. pp. 106

International Rice Research Institute-IRRI, 1976. Standard Evaluation System for Rice. IRRI, Los Baños. pp. 61

Ishiy, T., 1991. Desenvolvimento de genótipos de arroz (*Oryza sativa* L.) através de irradiação gama. Univeridade Estadual Paulista, Botucatu, SP. Tese de Doutiorado. pp. 67

Howeler, R.H., 1973. Iron-induced oranging disease of rice in relation to physical-chemical changes in a flooded oxisol. Soil Sci. Soc. of America Proceedings. **37**:898-903

Ponnamperuma, F.N., and R.U. Castro, 1972. Varietal differences in resistance to adverse soil conditions. In: IRRI. Rice Breeding, IRRI, Los Baños. p. 677-684

Syakudo, K. and H. Yamagata, 1960. Studies on artificial mutations in relation to plant breeding. VI. A preliminary analysis of genes responsible for several heading-period mutants caused by X-rays in rice. Jap. J. Breed. **10**:132

Webb, B.D., 1994. Current usage and anticipated changes in screening tests for evaluation cooking and processing characteristics of rice varieties and selections developed in United States rice breeding programs. Proc. Rice Technical Working Group **25**:43.

Improvement of tartary buckwheat by induced mutations with ^{60}Co gamma rays

Y. Tang[1], Z.X. Bin[2], and Z. Gang[1]
[1] *Xichang Agricultural College, Sichuan, 615013, P.R. China*
[2] *Sichuan Academy of Agri. Sciences, Chengdu, 610066, Sichuan, P.R. China*

Keywords: Fagopyrum tataricum, radiation mutagenesis, mutants, shattering resistance, flavone content, growth stimulation

Summary

The emergence rate, seedling height and growth of the root system of tartary buckwheat were progressively restrained through increased dose when seeds were treated with 100-800 Gy ^{60}Co gamma rays. Irradiation leads to very significant effects on emergence rate and seedling growth when seeds are treated with 700-800 Gy. However, when they are treated with 100-200 Gy, there is almost no effect. Seed irradiation can cause mutations of most traits of tartary buckwheat. Mutation frequency is very low, within 100-200 Gy, and continuously increases with doses from 300-600 Gy. The most suitable doses for induced mutations are 200-600 Gy. Mutagenic treatment alters flavone content and mutants with higher flavone content were selected from mutated populations. Furthermore, high shattering resistant mutants were also obtained and showed no segregation in the M_5. The mutant "5-2-62" has the best agronomic characters with high flavone content, high shattering resistance, appropriate growth duration and plant height. Its yields exceed the original variety by 57.7%. Results of regional trials show that the mutant "5-2-62" will be a new mutant variety which should be widely grown.

M. Maluszynski and K.J. Kasha (eds.),
Mutations, In Vitro and Molecular Techniques for Environmentally Sustainable Crop Improvement, 183–188.

Introduction

Buckwheat is not only the staple food of the minorities in remote upland regions of China, but is also a natural health food, which is rich in flavone so that more and more attention is being paid to it. Tartary buckwheat (*Fagopyrum tartaricum* (L.) Gaertn.) has been very low and unstable in yield. One of the reasons is that local varieties dominate buckwheat production, even though local varieties have good adaptability to unfavorable environments. However, their narrow genetic base limits the improvement of tartary buckwheat. Shattering, which restricts high yield, has been a difficult problem. A variety with high flavone content is needed for commercial production.

Tartary buckwheat is a self-pollinating crop and hybridization is very difficult due to small flower organs. At present, no success in crossing among tartary buckwheat varieties has been reported. Therefore, induced mutations are an important and effective breeding method for tartary buckwheat. Breeding with the use of induced mutations has been successful in common buckwheat (*Fagopyrum esculentum* Moench.) (Li, 1987; Wan, 1991). However, there have been no studies on induced mutation in tartary buckwheat. This paper reports on the use of gamma rays to improve flavone content and shattering resistance of tartary buckwheat.

Material and methods

Four tartary buckwheat varieties ('Jiujiang', 'Gela', 'Getu' and 'Geliuuche') with character differences and different origins were selected for the treatment.

Gamma ray treatment

Dry seeds with 14.2% water content were irradiated with 100, 200, 300, 400, 500, 600, 700 and 800 Gy of ^{60}Co gamma rays.

Planting and selection of M_1 to M_5 generation

Treated seeds were sown in pots. M_2 seeds from each M_1 plant were planted as a family in the field along with a parent variety. M_3, M_4 and M_5 were also planted on a plant family basis. The trait changes and variations in the M_2 generation were recorded. Selection began with the M_2 generation.

Determination of genetic stability of seed flavone content and shattering resistance

Mutants from Geliuuche were investigated for seed flavone content, shattering resistance and other important traits from M_2 to M_5 generations. Flavone was determined according to Tang and Gang (1981). Ten selected plants with consistent maturity were fixed on an oscillator and vibrated for 10 minutes to determine the number of shattered grains.

Yield trials

Four selected mutant lines with good agronomic traits were planted in regional trials of three counties in 1997. The plot size was 12 m^2 (4×3 m) with three replications. The best mutant was planted for production experiments in 1998, on a 3.3 ha field.

Results and discussion

Biological effects in M_1 generation

Effect of radiation on germination and emergence rate of seeds

Experimental results show that there is no obvious decrease in the germination of seeds compared to the check. Therefore, the germination rate of tartary buckwheat seeds was not affected by radiation doses up to 800 Gy. However, an increase in the dosage was accompanied by a progressive decrease in emergence rate (Figure 1).

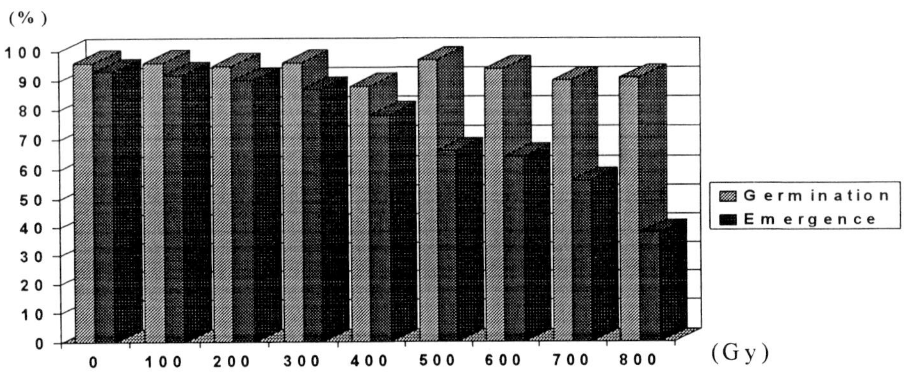

Figure 1. Effect of radiation with different dosage on germination and emergence rate

Effect of radiation on growth of seedling

In general, the growth of seedlings and their roots became slower with an increased dose (Table 1). This indicates that radiation leads to obvious restraining effects on root and seedling height. It was found that when seeds were treated with high radiation doses, not only the growth of the root was restrained, but the emergence and growth rate of lateral roots was slower. It is noteworthy that seedling height and root length increased to some extent when seeds were treated with 200 Gy gamma rays. It has been suggested that a low dose can stimulate plant growth (Sax, 1963).

Table 1. Effect of radiation on seedling height and root growth

Trait	Dose (Gy)								
	0	100	200	300	400	500	600	700	800
Seedling height (cm)	5.42	5.20	5.77	5.04	4.57	3.58	3.40	2.36	1.62
Main root length (cm)	4.21	4.56	6.48	5.75	53.77	2.48	1.38	0.98	0.00
No. of lateral roots	21.30	20.30	20.08	17.52	15.44	14.56	5.01	2.80	0.00

Variation of characters in the M_2 generation

Mutated characters were found in tartary buckwheat after treatment with various doses of gamma rays. Early or late, tall or dwarf, more or fewer numbers of grains/plant mutations were selected in M_2 populations (Table 2).

Table 2. Range of trait variations observed in the M_2 generation*

Trait	Dose (Gy)				
	0	300	400	500	600
Growth duration (days)	80-99	81-101	78-103	76-105	77-105
Plant height (cm)	70-102	61-105	52-102	45-101	49-100
No. of grains/plant	58-353	41-316	31-363	21-479	29-328
Grain weight/plant (g)	0.6-5.4	0.6-5.2	0.4-6.2	0.3-9.2	0.3-5.8
1000-grain weight (g)	11.5-22.4	11.0-22.8	9.8-24.7	10.3-23.6	10.6-27.3
No. of branches	3-10	1-11	2-10	1-11	2-9

*Data from four varieties: Geliuuche, Getu, Jiujiang and Gela

There were no significant changes of 1000-grain weight in progenies of 300 and 400 Gy treatments. M_2 populations from 500 and 600 Gy treatment showed a wider range of variation. Large grain mutants with 25-27 g of 1000-grain weight were selected in 600 Gy combinations of Geliuuche and Gela populations. This type of tartary buckwheat mutant seldom occurs in nature. Furthermore, radiation also caused alteration of grain shape and color. For example, the long cone changed into a peach shape and the black seed coat changed to brown.

Mutation frequency and dose

M_2 mutation frequencies in different doses of gamma ray treatment are listed in Table 3. It has been shown that although different varieties treated by different doses have some differences, they share the same tendencies. Mutation frequency is very low within 100-200 Gy and continuously increased with an increased dose within 300-600 Gy. But when the dose was more than 600 Gy, the mutation frequency decreased. M_2 mutation frequency for 800 Gy treatment could not be determined due to lack of seeds from the M_1. It can be concluded that the dosage for induced mutations ranges from 300 to 600 Gy of gamma rays.

Table 3. M_2 mutation frequency (%) of tartary buckwheat treated with different doses of gamma rays

Variety	Doses (Gy)						
	100	200	300	400	500	600	700
Geliuuche	0.03	2.87	10.26	18.29	23.72	25.89	24.74
Jiujian	0	0.65	6.61	13.16	18.65	24.08	26.52
Gela	0	1.42	7.33	15.42	20.71	20.91	17.81
Getu	0.01	2.09	8.14	18.10	24.48	25.46	27.02
Average	0.01	1.76	8.09	16.24	21.14	24.09	24.02

Flavone variation and its genetic stability

Flavone content of some mutants has been continuously determined from the M_2 over several generations. It has shown that radiation can increase the flavone content of tartary buckwheat (Table 4).

Table 4. Flavone contents of grains for some mutant plants, variety Geliuuche

Parent variety/ mutant	Flavone content (%)				
	Dose (Gy)	M_2	M_3	M_4	M_5
Geliuuche	0	2.29	2.40	2.34	2.31
3-2-6	300	3.15	2.98	3.07	3.11
4-2-12	400	2.86	2.73	2.97	2.68
4-2-19	400	2.74	2.94	2.90	2.82
5-72-3	500	3.36	3.46	3.45	3.37

There are no obvious differences in flavone content of the same mutants over the generations. The correlation coefficient of flavone content between the M_2 and M_3 was significant ($r=0.80$, $p=0.05$). This indicates that the inheritance of increased flavone content is very stable and that this desired character in tartary buckwheat can be improved by radiation.

Shattering resistance

Buckwheat is a crop with an indefinite racemose inflorescence. Long flowering and seed setting duration consist of two-thirds of the whole growth period. Dates of maturity are quite inconsistent on the same plant. When plants are harvested, the early seeds have shattered and therefore, the yields are greatly reduced. Breeding varieties with shattering resistance is an important approach to raising yields. To date, no tartary buckwheat varieties have shattering resistance.

The mutant "5-2-62" selected from ^{60}Co gamma treated Geliuuche populations has a very profound trait high shattering resistance, and this trait is inherited (Table 5).

Table 5. Comparison of grain shattering between the mutant "5-2-62" and original variety Geliuuche

Parent variety/		Shattering (%)			
mutant	Dose (Gy)	M_2	M_3	M_4	M_5
Geliuuche	0	39.8	40.7	45.9	37.4
5.2-62	500	4.6	6.2	3.8	5.1

Selection of desired mutants for yield trials

Selection was begun from M_2 and continued to the M_5 generation. Some selected mutants showed no segregation. The most outstanding mutants from irradiated populations of Geliuuche variety are listed in Table 6.

Table 6. Yield comparison among the desired mutants and the parental variety Geliuuche

Mutants	Yield (kg/ha)	Difference (kg/ha)
Geliuuche	1150.5	-
5-2-62	1809.0	+658.5**
4-2-12	1608.0	+457.5**
4-73-8	1242.0	+91.5
5-72-5	1200.0	+49.5
5-20-1	1167.0	+16.5
4-73-7	1033.5	-117.0*

* p=0.05,** p=0.01

The mutant 5.2-62 exceeds the parent variety in yield by 57.3% and has the best comprehensive traits with high seed setting, full grain and high shattering resistance. In April of 1997, regional trails with five varieties were made in three counties (Xidei, Yuexi and Meigu). The results show that the yields of the mutant 5-2-62 were the highest with an average yield of 2490 kg/ha, which is 33.1% higher than the local check variety Gela. In April of 1998, a large area-demonstration trial was carried out in Xidei county. The yields of the mutant line 5-2-62 average 2,575 kg/ha, which is 37.6% higher than the check Gela. Because of its excellent performance, this mutant variety will be applied as a new cultivar to cover a large area in China's tartary buckwheat growing regions.

References

Li, K., 1987. Inheriting breeding and good strain breeding of buckwheat. The Agric. Publishing House. **2**: 127-136

Sax, K., 1963. The stimulation of plant growth by ionizing radiation. Radiation Bot. **3**: 179-186

Tang, Y. and Z. Gang, 1981. Changes of flavone and rutin contents in buckwheat. Plant Physiol. **1**: 33-35

Wan, S., 1991. Contribution of inducing breeding to improvement of plant variety. The Agric. Publishing House. **2**: 57-58.

Evaluation of barley mutants for drought tolerance: a physiology - breeding approach

M.I. Cagirgan[1], J. Gorham[2], C. Toker[1], B. Uzun[1], K. Visser[3]
[1] *Department of Field Crops, Faculty of Agriculture, Akdeniz University, Antalya, Turkey*
[2] *Center for Arid Zone Studies, University of Wales-Bangor, Bangor, Gwynedd, Wales, U.K.*
[3] *Department of Plant Molecular Biotechnology, Center for Phytotechnology, RUL/TNO, Al Leiden, The Netherlands*

Keywords: Hordeum vulgare, abiotic stress tolerance, induced mutations, genetic male sterility, earliness, photosynthetic rates

Summary

Induced mutants are useful not only for cultivar development but also as research tools for basic studies. A barley mutant collection maintained at Akdeniz University was evaluated for development of cultivar and breeding populations for semi-dry conditions. Results in dry environments have indicated that both drought escape and drought tolerance can contribute to better grain yields under drought stress at heading and grain filling periods. Two types of mutants were investigated; those affecting traits involved in stress tolerance and those facilitating breeding procedures. Early-heading induced mutants, which escaped the drought period, had a greater grain yield because of a longer grain-filling period. Genetic male-sterile mutants were successfully used to overcome hand emasculations. Outcrossing frequency was determined on collected spikes from induced male sterile mutants and fluctuated greatly with environments and populations. However, it was sufficient for developing recurrent selection strategies. Using mutants as cross parents generated transgressive variability in yield components. The mutant collection was screened for characters tolerating abiotic stresses as well as its suitability for physiological and biochemical basic studies. An ABA insensitive induced mutant, Akdeniz-M-Q-54, was discovered. Photosynthetic rates for several mutants, including M-K-1, M-K-5, M-K-18 were consistently higher than the parent variety 'Kaya', both in field and greenhouse studies. Similarly, improvements were obtained for transpiration rate and stomatal conductance. Two glaucous mutants, M-K-49 (*sdw*) and M-K-55 had lower leaf Na^+ and Cl^- concentrations than Kaya when the leaves were sprayed with salt. M-K-49 had significantly higher leaf surface wax contents than Kaya at the vegetative stage.

M. Maluszynski and K.J. Kasha (eds.),
Mutations, In Vitro and Molecular Techniques for Environmentally Sustainable Crop Improvement, 189–199.

Introduction

Barley is an important crop especially for areas where drought and terminal heat stress are the most prevalent constraints, limiting its production and grain quality. Barley varieties that will produce stable and economic yield under these inconsistent rainfed conditions are needed. To accomplish this aim, barley germplasm should first be developed for the characters, which are closely associated with drought resistance (Sinha and Bansal, 1991).

Our main objective has been to characterize and use barley mutants for physiology research and breeding as well as to develop populations possessing high and stable yielding potential under drought and terminal heat stress conditions. The objective has been accomplished by combining the early heading and glaucous (high wax content) characteristics of certain mutants with other drought adaptation characteristics of local cultivars and land-races. This report will present the results of yield trials with the mutants and advanced lines, as well as summarize the results of physiological studies.

Material and methods

Mutant collection

Barley mutants (Table 1) were evaluated for direct and indirect use and also screened in physiological studies. All the mutants were selected from the two-row spring habit varieties, 'Kaya' and 'Quantum', after seed irradiation with gamma rays at 150 and 300 Gy and designated M-K and M-Q, respectively. The mutation spectrum includes such characters as heading date, spike morphology, plant height, glaucousness, genetic male sterility and anthocyanin-less (Cagirgan and Yildirim, 1990).

Selected lines and yield trials

The genetic male sterile mutants have been successfully used as females for cross breeding. Crosses were made in 1987 and 1989 in Tokat, Turkey using two crossing schemes: line x tester (Cagirgan, 1995) and a half-diallel scheme. F_2 populations were selected for several traits when grown in Tokat and a pedigree selection was applied in F_3 through F_5 for early heading, spike yield and/or its components at Aksu, Antalya - a potentially high yielding environment with an average of about 1000 mm rainfall annually. The selected F_6 lines were exposed to drought stress by late-spring sowing in 1993 in a semi-dry environment with about 429 mm rainfall annually (in Urkutlu, an upland of Antalya) and evaluated for grain yield. The selection for grain yield was continued in the F_7. Final selections were grown in a Randomised Complete Blocks Design (three replications) at two different sowing times in 1995 (Exps. A10-2 and A11-2). A plot consisted of six rows, each 4 m long. The seeding rate was 400 grains/m^2. Sixty kg P_2O_5 and N per hectare were applied. The following characters were observed: biomass yield and grain yield; harvest index; days to heading after 1st May; number of spikes/m^2; number of kernels/spike; 1000-kernel weight, and plant height. Analysis of variance (ANOVA) was applied to all characters studied and means were compared using Duncan's Multiple Range Test.

In a separate set of experiments grown in 1993 and 1994 at two locations, University Campus and Aksu, base and selected sub-populations segregating for male sterility, spike yield and early heading were evaluated. Similar agronomic practices and design were applied with four replications. The above-described characters were also measured, except for the number of spikes and plant height. Spike yield was calculated on the basis of kernels per spike and 1000 kernels weight. The data were analyzed using MSTAT-C software.

Table 1. Gamma-ray induced mutants collection in two-row spring barley varieties Kaya (M-K) and Quantum (M-Q) maintained at Akdeniz University, Antalya, Turkey

Basic characteristics of the mutant category	Mutants in the category and additional characteristics
Early-heading and anthocyaninless	M-K-1, M-K-2, M-K-3, M-K-4, M-K-5, M-K-14, M-K-15, M-K-16, M-K-17, M-K-18, M-K-19, M-K-20, M-K-25, M-K-26, M-K-27, M-K-28, M-K-29, M-K-32, M-K-33, M-K-34, M-K-44, M-K-45, M-K-51, M-K-64, M-K-65, M-K-66, M-K-67, M-K-81, M-K-84, M-K-88
Glacousness	M-K-49 (salt Tolerant, *Sdw*, More Wax)
	M-K-52 (*sdw*,more wax)
	M-K-46, M-K-47, M-K-48, M-K-55, M-K-56 (more wax)
	M-K-57,
	M-K-68 (less-wax, pale-green, tall, late flowering)
	M-Q-76 (*eceriferum*, waxless, semifertile, late flowering)
ABA-insensitive	M-Q-54 (tall, higher biomass yield, giant)
Genetic male-sterility	M-K-42, M-K-43, M-Q-87
	M-K-89 (6-row), M-Q-78 (irregulare)
Six-row	M-K-36, M-K-37, M-K-38, M-K-59, M-K-61, M-K-62, M-K-63
	M-K-59 (intermedium)
Compact spike and anthocyaninless	M-K-8, M-K-9 (earlier in heading, smooth awn)
	M-K-10, M-K-11 (smooth awn)
	M-K-23 (smooth awn) m-k-24 (normal awn), isogenics
	M-K-83 (smooth awn, long grain)
	M-K-85 (oval grain shape)
Compact spike (with normal anthocyanin pigmentation)	M-K-6, M-K-7 (dark green), M-K-13, M-K-30, M-K-53
	M-K-39 (multinoded, seed not available)
	M-K-40 (dwarf, seed not available)
	M-K-41, M-K-58, M-K-82 (semidwarf)
	M-Q-73, M-Q-74, M-Q-75 (ert, large grain, smooth awn)
	M-Q-80 (ert, semidwarf)
Lax spike	M-K-35 (semidwarf, late in heading)
Triple-eared	M-Q-86 (seed not available)
Naked grain	M-K-31
	M-K-57 (light green)

Physiological investigations

The mutants varying in plant height were screened for sensitivity to gibberellic acid (GA_3) and ABA in germination medium. One mutant, M-Q-54, taller than the wild-type, was studied further since isolated embryos were able to germinate in the presence of 10^{-4} M ABA. Germinated embryos were grown in a phytotron at the Center for Phytotechnology, Leiden University, The Netherlands. The details for germination tests, ABA sensitivity test, RNA isolation and analysis and ABA measurements were reported by Visser *et al.* (1996).

Photosynthetic gas exchange parameters (carbon dioxide exchange rate, transpiration, stomatal conductance and internal CO_2 concentration) were measured with a CIRAS-1 infra-red gas analyser (PP Systems) at saturating light intensities (about 1500 μmol m^{-2} s^{-1} PAR). Air supplied to the chamber was dry and had a CO_2 concentration of 350 ppm. For field measurements a standard broad-leaf chamber was used with natural sunlight, whereas an automated broad-leaf chamber was used at 1500 μmol m^{-2} s^{-1} PAR and ambient temperature for greenhouse work. Leaf temperatures in the field were measured with a Calex MS2 infra-red thermometer. Inorganic cations in leaf sap were determined by ion chromatography on a Dionex CS12 column eluted at 50°C with 25 mol m^{-3} methanesulphonic acid.

Leaf contact angles of 5 μl droplets of water were measured using a video camera connected to a Wild binocular microscope mounted horizontally. The leaf was attached to a glass slide with acrylate glue, and illuminated with a fibre-optic light source. Gravimetric wax analysis followed extraction of epicuticular wax with 1:1 chloroform: heptane for 30 s. The extract was reduced in volume on a rotary evaporator, filtered (0.45 μm) in a pre-weighed microcentrifuge tube and dried before weighing.

Results and discussion

Mutant collection

The mutant collection has been intensively evaluated for mutations of quantitative traits and their heritability determined (ranging between 0.20 and 0.94). Most mutants were characterized by much earlier heading, as much as 20 days. Thousand kernels weight of certain mutants increased by 10 g, as a result of a longer grain filling period and decreased number of seeds/spike. Grain yields of mutants were lower than the control under normal yielding environments but higher under drought stress (Cagirgan and Yildirim, 1990). Mutants were grouped according to their usefulness for developing germplasm for drought conditions and for researching drought physiology (Table 1).

In the Kaya background, thirty early heading mutants were selected that behaved as such consistently over contrasting environments. Lower rainfall and its most erratic distribution during heading and grain filling periods necessitates the development of barley cultivars that can escape from drought periods by heading earlier, and thus taking advantage of available water in the soil. This mechanism has been shown to work with early heading mutants of locally adapted germplasm (Yildirim and Cagirgan, 1990). Among mutations, early heading types are not rare, since the earliness often was observed in connection with other changes in morphological and physiological characters (Ukai and Yamashita, 1981; Micke *et al.*, 1990).

Seven mutants in the Kaya background were glaucous to various degrees. These mutants were identified as M-K-46, M-K-47, M-K-48, M-K-49 (sdw), M-K-52 (sdw), M-K-55, and M-K-56. The ability of barley plant to survive severe drought conditions depends on its ability to restrict water loss through the leaf epidermis after stomata attain minimum aperture. A thick waxy cuticle, therefore, is advantageous in reducing water loss as a dehydration avoidance mechanism (Blum, 1988).

Five male-sterile plants, characterized by open glumes at flowering, were identified in the M_2 and M_3 populations of the Kaya and Quantum varieties and designed as M-K-42, M-K-43, M-K-89, M-Q-78, and M-Q-87. These mutants have been used to facilitate crossing and to select more turgid plants, which are more prone to out-crossing.

Selected lines and yield trials

In experiment A 10-2 (Tables 2 and 3), biomass yield differences were not significant but 'Kilinc', a local land-race, and some selected lines (Nos. 3,4,9,12 and 13) looked promising. Entry No.12 produced the highest grain yield, 2.27 t/ha, but again there were few significant differences. However, it is important to note that these lines were earlier heading than their parents and generally performed well in most traits measured. The early mutant M-K-88 performed well for most traits. Some lines with significantly improved harvest index were also obtained (e.g. Nos. 8, 7; early-heading mutant M-K-88). There was no statistically significant difference for number of spikes per m^2 ($p<0.05$) among the entries tested. Entry No.5 and Kilinc had the highest value for 1000-kernel weight (43.4 g). Plant height ranged between 53-88 cm among the entries and tended to be higher than the parents with thirteen selections being significantly taller.

Table 2. Biomass, grain yield, harvest index and days to heading of selected barley mutant lines grown in 1995, Urkutlu (Experiment A10-2)

Entry No.	Name	Biomass (t/ha)	Grain yield (t/ha)	Harvest index (%)	Days to heading
1	87K-6	6.32 ab	1.67 abc	26.1 efg	23.7 cd
2	87K-9	5.58 ab	1.82ab	32.6 abc	23.3 de
3	87K-12	6.56 a	2.12 a	32.4 abc	22.7 de
4	87K-13	6.72 a	2.02 ab	30.0 abcdef	19.3 fg
5	87K-231	5.83 ab	1.48 abc	25.7 fg	24.3 bcd
6	87E-349	5.31 ab	1.82 ab	34.3 ab	14.7 hijk
7	87E-356	6.08 ab	2.14 a	34.9 a	14.7 hijk
8	87E-369	6.16 ab	2.18 a	35.0 a	19.3 fg
9	87E-396	6.64 a	1.94 ab	28.5 bcdef	11.0 l
10	87E-408	6.26 ab	1.97 ab	31.1 abcdef	15.0 hijk
11	87E-413	6.30 ab	2.05 ab	31.9 abcd	13.7 jkl
12	87E-418	6.64 a	2.26 a	34.0 ab	18.0 fgh
13	87E-424	6.60 a	2.05 ab	31.2 abcdef	12.7 kl
14	87E-425	6.13 ab	1.79 ab	29.3 abcdef	14.0 ijkl
15	89Y-373	5.36 ab	1.71 abc	30.5 abcdef	16.7 ghij
16	89Y-375	5.83 ab	1.88 ab	32.2 abc	20.3 ef
17	89Y-385	5.11 ab	1.59 abc	30.8 abcdef	15.3 hijk
18	89Y-394	5.55 ab	1.67 abc	29.9 abcdef	17.3 fghi
19	89Y-398	5.59 ab	1.80 ab	31.9 abcd	18.0 fgh
20	Kaya	5.01 ab	1.30 bc	26.2 defg	26.7 abc
21	Quantum	5.42 ab	1.52 abc	27.5 cdef	27.3 ab
22	M-K-88	6.16 ab	2.14 a	34.8 a	13.3 jkl
23	Tokak	5.52 ab	1.45 abc	26.2 defg	29.7 a
24	Ankara	4.43 b	0.96 c	22.1 g	29.3 a
25	(Kilinc)	6.90 a	2.17 a	31.6 abcde	24.0 cd

Means followed by the same letter are not statistically different (Duncan's Multiple Range Test, $p<0.05$)

Table 3. Number of spikes, grains per spike, 1000-grain weight and plant height of selected barley mutant lines grown in Urkutlu (Experiment A10-2)

Entry No.	Name	Spikes/m^2	Kernals/spike	1000-kernal weight (g)	Plant height (cm)
1	87K-6	84 bcd	27.5 a	38.8 abc	73 cdefghi
2	87K-9	90 abcd	22.3 efgh	40.8 ab	70 efghi
3	87K-12	109 abcd	23.0 defgh	35.3 cd	82 abc
4	87K-13	105 abcd	27.3 ab	37.4 bcd	85 ab
5	87K-231	103 abcd	21.4 h	43.4 a	67 ghij
6	87E-349	110 abcd	21.3 h	39.3 abc	71 defghi
7	87E-356	105 abcd	22.9 defgh	39.3 abc	77 bcdefg
8	87E-369	104 abcd	23.9 cdeefgh	39.0 abc	83 abc
9	87E-396	116 abcd	18.4 i	39.0 abcd	81 abc
10	87E-408	116 abcd	23.8 cdefgh	36.0 bcd	77 bcdefg
11	87E-413	127 ab	21.7 fgh	37.6 bcd	75 cdefg
12	87E-418	115 abcd	24.6 bcdefg	36.8 bcd	88 a
13	87E-424	110 abcd	21.9 fgh	35.7 bcd	79 abcde
14	87E-425	109 abcd	21.7 gh	39.3 abc	82 abc
15	89Y-373	78 cd	23.7 cdefgh	34.3 cde	69 fghij
16	89Y-375	114 abcd	25.3 abcde	32.9 de	78 abcdef
17	89Y-385	118 abc	21.5 gh	37.1 bcd	74 cdefgh
18	89Y-394	113 abcd	23.5 defgh	36.5 bcd	75 bcdefg
19	89Y-398	96 abcd	25.5 abcd	36.7 bcd	77 bcdefg
20	Kaya	105 abcd	26.6 abc	29.9 e	65 hij
21	Quantum	68 d	24.7 abcdef	39.4 abc	64 ij
22	M-K-88	135 a	23.3 defgh	35.7 bcd	81 abcd
23	Tokak	110 abcd	22.2 fgh	37.2 bcd	60 jk
24	Ankara	68 d	24.6 abcdefg	35.9 bcd	53 k
25	Kilinc)	95 abcd	22.5 efgh	43.4 a	74 cdefgh

Means followed by the same letter are not statistically different (Duncan's Multiple Range Test, $p<0.05$)

In Experiment A11-2 (Tables 4 and 5), there was more variability among lines and the early heading mutant, M-K-88, had the highest values for biomass, grain yield and harvest index and lowest value for days to heading. Some lines were also better than the control for 1000-kernel weight.

Determination of the traits related to drought resistance as selection criteria has been a major problem for establishing efficient breeding programs (Kronstad and Moss, 1991). Although a range of traits has been suggested to improve yield capacity under drought stress, the use of physiological traits has not often been used by breeders. Additionally, most breeding programs conduct selection and germplasm evaluation for drought under the more favourable conditions on research stations. Therefore the selected materials may be poorly adapted to farmer's fields. Evaluation and selection of breeding materials in the farmers' fields may give a better answer, as to which entries and crosses are more suited for cultivation under drought conditions.

Genetic diversity is an important necessity for success in plant breeding. Induced mutations play an important role in germplasm enhancement and thus help to avoid genetic vulnerability. Many mutants that have been induced are, however, of very restricted direct use as they are often associated with unwanted changes in other characters. This may result from pleiotropic effects of the mutated gene(s) or additional mutational events in an individual plant. Despite this, it is well documented that different alleles mutated in a genetic background can interact with each other and with naturally existing alleles when transferred into a different genetic background (Gaul, 1964). As simply stated by Ramage (1987) every gene or gene complex affecting a trait has a "happy home", or a background genotype that is suitable for the desired optimum expression of that gene or gene complex. In this case, it is also very possible to obtain transgression in economical characters as well as more desirable character combinations such as e.g. lodging resistance and earliness.

Table 4. Biomass, grain yield, harvest index and days to heading of selected barley mutant lines grown in Urkutlu (Experiment A11-2)

Entry No.	Entry Name	Biomass (t/ha)	Grain yield (t/ha)	Harvest index (%)	Days to heading
1	87Y-14	5.01 cd	1.62 de	32.3 bcd	23.0 ef
2	87Y-18	5.18 cd	1.19 ef	23.1 f	27.7 abc
3	87Y-42	5.43 bcd	1.32 ef	24.0 ef	29.3 a
4	87Y-44	5.21 cd	1.69 de	32.2 bcd	25.0 cde
5	87Y-73	6.51 abc	2.40 abc	36.8 ab	24.3 def
6	87Y-77	2.98 e	0.89 f	29.3 d	15.7 h
7	87Y-105	7.69 a	2.26 bcd	29.5 d	21.7 f
8	87Y-162	6.29 abcd	1.86 cde	29.6 d	22.7 ef
9	87Y-296	7.64 a	2.74 ab	35.7 abc	22.7 ef
10	87Y-316	7.02 ab	2.47 abc	35.0 abc	19.0 g
11	87Y-322	7.02 ab	2.61 ab	37.2 ab	13.3 h
12	87Y-344	6.89 ab	2.59 ab	37.5 ab	14.3 h
13	Y-74	5.03 cd	1.72 de	33.7 abcd	28.3 ab
14	Y-106	5.44 bcd	2.05 bcd	37.3 ab	26.3 bcd
15	Y-80	4.68 d	1.35 ef	28.7 de	27.0 abcd
16	Quantum	7.20 a	2.49 abc	34.3 abcd	28.0 ab
17	Kaya	5.56 bcd	1.70 de	30.7 cd	26.7 abcd
18	M-K-88	7.74 a	2.98 a	38.3 a	13.0 h

Means followed by the same letter are not statistically different (Duncan's Multiple Range Test, $p<0.05$).

Male sterility facilitated recurrent selection (MSFRS) could be used to incorporate desirable agronomic characters into landraces without changing its adaptation and narrowing the genetic base of the landrace to any appreciable degree (Falk, 1989). Application of recurrent selection schemes have not been extensively used to improve self-pollinating crops because of the difficulty in obtaining large numbers of crosses which are necessary for each selection cycle. Genetic male-sterile genes however, have been used effectively to eliminate hand emasculation in plant breeding programs, and thus to facilitate the development of recurrent selection populations in self-pollinated crops. A high level of natural outcrossing on male sterile plants is desirable for recurrent selection strategies to avoid labour-intensive hand-pollination. Spontaneous or induced male sterility in locally adapted germplasm may provide a better source of male sterility than exotic germplasm. The level of outcrossing of three male sterile facilitated recurrent selection populations (MSFRSPs) was studied in 1991 and 1992 under two adverse environmental conditions: Pullman, Washington, USA and Antalya, Turkey. The seed set, determined on collected spikes from male sterile plants, fluctuated greatly (4.3-21.1%) from location to location, year to year and population to population (Cagirgan and Ullrich, 1991).

Table 5. Number of spikes, grains per spike, 1000-grain weight and plant height of selected barley mutant lines grown in Urkutlu (Experiment A11-2)

Entry No	Entry Name	Spikes/m^2	Kernals/spike	1000-kernal weight (g)	Plant height (cm)
1	87Y-14	110 ab	25.2 abc	38.8 def	70 cd
2	87Y-18	78 b	23.3 cde	42.8 abc	69 cde
3	87Y-42	79 b	24.7 abc	44.8 ab	67 de
4	87Y-44	87 ab	23.9 bcd	45.2 a	65 def
5	87Y-73	93 ab	23.8 bcd	41.8 abcd	67 de
6	87Y-77	118 ab	19.5 g	30.0 j	46 h
7	87Y-105	113 ab	26.4 a	41.3 bcd	83 a
8	87Y-162	98 ab	21.6 defg	44.7 ab	83 a
9	87Y-296	113 ab	26.7 a	37.3 efg	84 a
10	87Y-316	118 ab	24.5 abc	37.5 efg	82 a
11	87Y-322	117 ab	19.9 fg	43.7 abc	77 abc
12	87Y-344	126 ab	21.2 efg	45.2 a	79 ab
13	Y-74	134 a	23.3 cde	31.7 ij	56 g
14	Y-106	118 ab	22.0 def	33.0 hij	58 fg
15	Y-80	106 ab	26.6 a	34.7 ghi	61 efg
16	Quantum	113 ab	25.0 abc	40.5 cde	68 de
17	Kaya	133 a	26.1 ab	33.5 hij	73 bcd
18	M-K-88	122 ab	21.3 efg	36.0 fgh	82 a

Means followed by the same letter are not statistically different (Duncan's Multiple Range Test, $p<0.05$).

Physiological studies

A cooperative project between Akdeniz University and Leiden University to discern the role of gibberellic acid (GA_3) and ABA in controlling the germination of barley seed was started in 1993. Mutant lines with various plant heights (Table 1) were tested for sensitivity to GA_3 and ABA. No difference in response to GA_3 for germination was found among the mutants tested. However, the dissected embryos of the M-Q-54 mutant germinated at least 10 h earlier than those of the parent Quantum in the presence of 10^{-4} M ABA. Consequently, this mutant line was selected for further studies. M-Q-54 was selected in the M_2 generation of Quantum barley for its erect habit, taller plant height and more vigorous growth than the parent variety (Cagirgan *et al.*, 1995). These attributes were confirmed in the M_3 and M_4 generations grown in two replicated progeny rows. It was also revealed that the mutant has better cold tolerance and wider adaptability especially in areas where both soil fertility and rainfall are limiting factors. Because of its longer basal internode it is susceptible to lodging in high-yielding environments.

In our study of the role of abscisic acid (ABA) in germination of barley seed, it was established that the mutant M-Q-54 was also ABA-insensitive. The mutant was selected by assaying germination in the presence of 10^{-4} M ABA. Dissected embryos of the mutant germinated at least 10 h earlier than those of the parent variety. The concentration of 5 x 10^{-4} M ABA was required for 50% germination inhibition of the mutant and 1 x 10^{-6} for the parent. The 50% inhibition of an ABA-induced *Rab* gene was 5 x 10^{-6} M in isolated embryos of both the mutant and its parent. This result suggests that ABA signal transduction pathways were not affected in the mutant. When isolated embryos were imbibed in water, ABA was released from the mutant and the parent embryos at the same rate. However, the free ABA level in the incubation medium of the mutant showed a much faster decrease than that of the parent variety.

The results suggest that turnover of ABA outside the embryo is a determining factor in germination of barley seeds (Visser *et al.*, 1996). A mutant with low basal level of ABA has been previously reported by Walker-Simmons *et al.* (1989). The mutant M-Q-54 was released in August 1997 as a germplasm to study the role of ABA in seed germination and was registrated by Crop Science Society of America, under the name Akdeniz-M-Q-54 - Reg. No.GP-134, PI 599301 (Cagirgan *et al.*, 1998).

Another co-operative study to screen the mutant collection for characters related to stress tolerance has been established between Akdeniz University and University of Wales-Bangor in 1997. The photosynthetic capacity of flag leaves of barley mutants was assessed in the field at Ürkütlü in Antalya in May 1998 and in a greenhouse in Bangor in July 1998 (Figure 1). Photosynthetic rates for several mutants, including M-K-1, M-K-5, M-K-18, were consistently higher than rates for Kaya, while for other lines the data were inconsistent. Absolute rates in the cooler and darker conditions in Bangor were lower than those obtained in the field in Anatalya. Similar trends were observed for transpiration rate and stomatal conductance. There were no significant differences in leaf temperatures or cation contents in Ürkütlü.

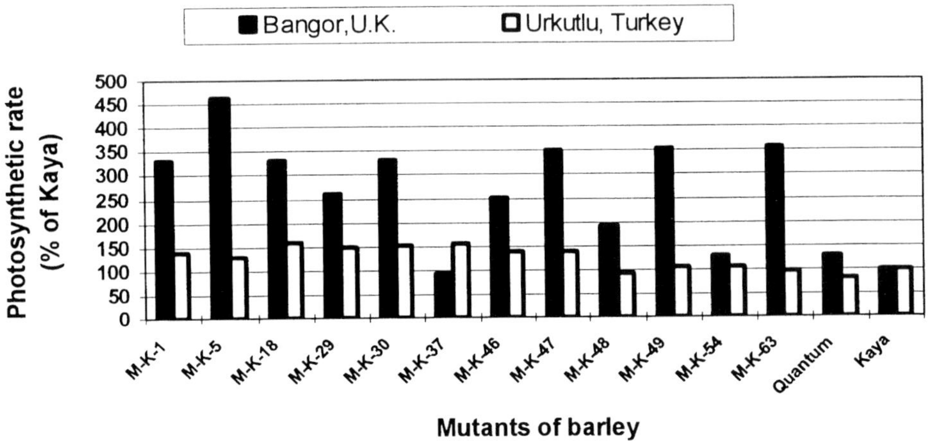

Figure 1. Photosynthetic rates of mutants from parent variety Kaya, in the field in Turkey and in the greenhouse in Wales

The performance of some waxy mutants in relation to foliar salt absorption is shown in Table 6. Only M-K-49 had significantly higher leaf surface wax contents than Kaya at the vegetative stage (designation as a waxy mutant is based on glaucousness at the reproductive stage), and none of the mutants had lower droplet contact angles (which would indicate a more hydrophobic leaf surface) than the parent. However, both M-K-49 and M-K-55 had lower leaf Na^+ and Cl^- concentrations than Kaya after the leaves were sprayed with salt. Thus the extent of foliar absorption of salt is not directly related to the hydrophobicity of the leaf surface, but may be associated with changes in a specific wax fraction. In separate experiment the uptake of salt by roots was similar in Kaya and the waxy mutants (data not shown).

Table 6. Leaf surface properties and foliar salt absorption in Kaya and Kaya waxy mutants

Line	Contact angle (°)	Epicuticular wax (mg/g FW)	Leaf Na^+ (mol m^{-3} sap)	Leaf Cl^- (mol m^{-3} sap)
Kaya	65.6 ± 4.7	11.8 ± 1.3	73.8 ± 7.6	220 ± 32
M-K-48	72.5 ± 3.0	10.4 ± 2.1	67.4 ± 7.0	224 ± 25
M-K-49	79.4 ± 2.7	15.4 ± 3.0	44.0 ± 4.5	171 ± 23
M-K-55	86.3 ± 2.9	10.8 ± 2.6	32.6 ± 3.5	144 ± 16

Acknowledgements

The authors acknowledge the supports of the IAEA under Research Agreement No: 8637 and British Council, Ankara under the Link Project Akdeniz/Wales-Bangor. Data in Table 6 was provided by Syed Ali Rehman.

References

Blum, A. (1988) Plant Breeding for Stress Environments. CRC Press, Boca Raton, Florida

Cagirgan, M.I., 1995. Modification of barley mutants differing in heading time by changing the genetic background: line x tester analysis for yield and yield components. In: Induced Mutations and Molecular Techniques for Crop Improvement. IAEA, Vienna. pp.698-703

Cagirgan, M.I. and S.E. Ullrich, 1991. Male sterile facilitated recurrent selection - A review. In: ESNA XXIInd Annual Meeting. Antalya, Turkey. pp.53 (Abstract)

Cagirgan, M.I., K. Visser, C. Toker, M. Wang, M.B. Yildirim, M.E. Tugay, and F. Heidekamp, 1998. Registration of M-Q-54, a fast-germinating ABA-Insensitive Mutant of 'Quantum' barley. Crop Sci. **38**: 1410

Cagirgan, M.I. and M.B. Yildirim, 1990. Macromutational variability in metric traits of barley. Ak.Ü.Zir.Fak.Derg. **3**: 139-152

Cagirgan, M.I., M.B. Yildirim, and M.E. Tugay, 1995. A giganteum mutant in barley. Barley Genet.Newsl. **25**: 21-22

Falk, D.E., 1989. Male sterile faciliated recurent selection. In: Current Options for Cereal Improvement. Maluszynski, M. (Ed.) Kluwer Acad.Publ., Dordrecht. pp.31-38

Gaul, H., 1964. Mutations in plant breeding. Radiation Bot. **4**: 155-232

Kronstad, W.E. and D.N. Moss, 1991. Requirements of plant breeders from physiologists to increase selection efficiency. In: Physiology-Breeding of Winter Cereals for Stressed Mediterranean Environments. INRA, Montpellier. pp.319-335

Micke, A., B. Donini, and M. Maluszynski, 1990. Induced mutations for crop improvement. Mutat.Breed.Rev. **7**: 1-41

Ramage, R.T., 1987. A history of barley breeding methods. Plant Breed.Rev. **5**: 95-138

Sinha, S.K. and K. C. Bansal, 1991. Evaluation of drought tolerance of *Triticum aestivum* and related species under field conditions. In: Physiology-Breeding of Winter Cereals for Stressed Mediterranean Environments. INRA, Montpellier. pp.254-269

Ukai, Y. and A. Yamashita, 1981. Early maturing mutants induced by ionizing radiations and chemicals in barley. In: Barley Genetics IV. Edingburgh. pp.846-856

Visser, K., A. Vissers, M.I. Cagirgan, J. Kijne, and M. Wang, 1996. Rapid germination of a barley mutant is correlated with rapid turnover of abscisic acid outside the embryo. Plant Physiol. **111**: 1127-1133

Walker-Simmons, M., D.A. Kurdna, and R.L. Warner, 1989. Reduced accumulation of ABA during water stress in a molybdenum cofactor mutant of barley. Plant Physiol. **90**: 728-733

Yildirim, M.B. and M.I. Cagirgan, 1990. Improvement in the grain yield capacity by mutagenesis in barley. In: Contribution of Plant Mutation Breeding to Crop Improvement Vol. 2. IAEA, Vienna. pp.401-405.

Mutations affecting nodulation in grain legumes and their potential in sustainable cropping systems

C.R. Bhatia, K. Nichterlein and M. Maluszynski
Plant Breeding and Genetics Section, Joint FAO/IAEA Division, P.O. Box 100, Vienna, Austria

Key words: inheritance, nodulation mutants, nitrogen fixation, symbiotic interaction

Summary

Many spontaneous and a large number of induced mutants that show altered nodulation pattern have been isolated in pea, soybean, common bean, faba bean, chickpea, groundnut and pigeonpea. Available information on nodulation mutants in these crops is summarised. The importance of nodulation mutants in basic studies on plant-microbe symbiotic interactions, nitrogen fixation and breeding of cultivars with higher yield and nitrogen fixation rate are examined. The nodulation mutants, after inoculation with specific bacterial strains or a number of different strains, show either: no nodulation (nod-), few nodules (nod+/-), ineffective nodulation (fix-), hyper nodulation (nod++) or hypernodulation even in the presence of otherwise inhibitory nitrate levels (nts). No spontaneous hypernodulation or nts mutants have been found, all have been induced in independent experiments using different cultivars of pea, soybean and common bean after mutagenising seeds. Most nodulation mutants show monogenic recessive inheritance, though semi-dominant and dominant inheritance is also reported. Nodule number is controlled by a process known as autoregulation; hypernodulating mutants show relaxed autoregulation. By grafting shoots of hypernodulating soybean mutant on normal nodulating soybean, mungbean and hyacinth bean, presence of a common, translocatable signal has been shown. Nodulation mutants have contributed to the understanding of the genetic regulation of host-symbiont interactions, nodule development and N fixation. Initially, the hypernodulating mutants were found to be poor in yield. Using the induced hypernodulating mutant, a new soybean cultivar 'Nitrobean 60', has been released in Australia. This cultivar is reported to have given 15% higher yield over cv. 'Bragg,' and contributed a higher amount of fixed N to the following cereal crop in rotation. Prospects of using the nodulation mutants in developing grain legume cultivars that combine high yield with high residual N, within the bioenergetic constraints, for developing sustainable cropping systems are examined.

M. Maluszynski and K.J. Kasha (eds.),
Mutations, In Vitro and Molecular Techniques for Environmentally Sustainable Crop Improvement, 201–226.

Introduction

Grain legumes are an important group of crops valued for their seeds with high protein content. Soybean and groundnut are also an important source of vegetable oils. These crops are valued for their ability to grow in soils poor in nitrogen (N), and ability to meet part of their N requirements through the fixation of atmospheric N to ammonium N that can be used by the plants. Biological nitrogen fixation (BNF) by the leguminous crops, is an important component of the nitrogen cycle in agro-ecosystems. Traditionally, legumes formed a component of pastureland vegetation in ley (crop - pasture rotations) farming or were grown as green manure crops, to restore soil fertility. Such farming systems were sustained over a long period, though the overall productivity was much lower in comparison to the present yield levels of cereals harvested with the application of chemical fertilizers. With increasing population, and decreased availability of cultivable land, the practice of ley farming, and cultivation of green manure crops has become uneconomical in densely populated countries (Ali, 1999). Integration of grain legumes into the cropping systems has been adopted in many regions. This gives better economic returns to the farmers, as well as contributing towards maintaining soil fertility. Breeding of legumes for enhanced symbiotic N_2 fixation was recommended by an expert group (Anon., 1984). The importance of legumes, and particularly the grain legumes, in sustainable agriculture have been the subject of many international conferences (e.g., Nitrogen Fixation International Congresses, 1970-1999; Nitrogen Fixation for Sustainable Agriculture, 1994; Sustainable Agriculture for the Tropics, 1995), and research initiatives (NATO Advanced Research Workshop, 1996; FAO-IAEA programmes in the recent past (Anon., 1994; Anon., 1997a).

In the long term, sustainable land productivity, and higher crop yields can only be maintained by replenishing the soil nutrients removed by the crop (Peoples *et al.*, 1994). In plant growth and development, N acquisition and assimilation is only next to photosynthesis in importance (Vance, 1997). The two processes are inter linked and determine crop productivity. Thus, N fixation by the grain legumes, and their soil N saving effect, can contribute a great deal towards evolving sustainable, productive, cropping systems. Large increases in crop yield can be realised on a sustainable basis, without substantial increases in external inputs, by enhancing the nitrogen fixation by the grain legumes (Danso, 1995; Fried, 1995; Herridge *et al.*, 1995b; Boddey *et al.*, 1997). Dobereiner (1997) has reported that soybean selected for zero N application in Brazil has become the country's largest export product.

Extensive studies on plant-microbe interactions, based on bacterial and host plant mutants, have established the predominant control of the legume plant in the establishment and functioning of the symbiosis. An inventory of the symbiotic legume mutants, then available, was included in the review by Caetano-Anolles and Gresshoff (1991). Twenty-two nodulation mutants in pea, eight in soybean, five in chickpea and two in groundnut were reported. Gresshoff et al. (1999) pointed out that in recent years a large number of nodulation mutants have been identified in different crops (Tables 2-4). Induced mutations have been widely used for the genetic enhancement of grain legumes (Micke, 1988) and 265 new cultivars using this approach have been released for commercial cultivation (Bhatia *et al.*, 1999). However, most of the nodulation mutants have been obtained by scientists with primary interests in BNF. N fixation and nodulation traits have been largely neglected by the grain legume breeders (Fried, 1995;

Herridge and Danso, 1995a). In this paper current information on spontaneous and induced nodulation mutations in different grain legumes is compiled. The importance of nodulation mutants in basic studies on legume symbiosis, nitrogen fixation and breeding of cultivars with higher yield and nitrogen fixation are examined. The authors hope that this publication will stimulate searches for nodulation and root mutants in the available mutant collections, and in new mutation experiments with grain legume crops.

Nitrogen fixation and nodulation in grain legumes

Fixation of atmospheric N by the leguminous plants is an extremely complex process that has been extensively investigated, yet its understanding is inadequate to derive full benefits for agriculture. Comprehensive reviews on different aspects of the subject are available (Djordjevic et al., 1987; Long, 1989; Nap and bisseling, 1990; Sprent and Sprent, 1990; Brevin, 1991; Caetano-Anolles and Gresshof, 1991; Denaire *et al.*, 1992; Gresshoff, 1993). Vance (1983) has referred to *Rhizobium* infection and nodulation as a beneficial plant disease. Djordjevic et al. (1987), drawing similarities between *Agrobacterium* and *Rhizobium* infections, have referred to the latter as the refined parasite of the legumes. Symbiotic N fixation involves gene-controlled interactions between the host plant and the *Rhizobium, Bradyrhizobium, Sinorhizobium* or *Azorhizobium* species of soil bacteria collectively known as "rhizobia". In response to a variety of substances excreted by the host plant roots, specific soil bacteria are attracted to a particular legume species, leading to the formation of specialized root organ, the nodule. The steps involved in establishment of successful symbiosis are: multiplication and colonization at the root surface (*Roc* **ro**ot **c**olonization), adhesion of bacteria to the root hair surface (*Roa* **ro**ot **a**dhesion), curling or branching of host root hairs (*Hab* and *Hac* (root **ha**ir **b**ranching and **c**urling), formation of **inf**ection thread (*Inf*), induction of meristems in the host roots for the **no**dule **i**nitiation and differentiation (*Noi*), **ba**cterioid **r**elease from the infection thread (*Bar*), **ba**cterial **d**ifferentiation (*Bad*), onset of **ni**trogen **f**ixation (*Nif*), **no**dule function **p**ersistance/maintenance (*Nop*), development of **co**mplementary **f**unctions associated with N fixation (*Cof*) and its transport (Sprent and Sprent, 1990; Caetano-Anolles and Gresshoff, 1991).

Estimates of the amount of nitrogen fixed

Though it is a common observation that cereal or other non-legume crops grown in plots after a grain legume crop show better growth and yield, precise estimates of the amount of N fixed, and beneficial effects on the following cereal crop vary considerably. The range of variation reported is given in Table 1. Physical, environmental, nutritional and biological parameters, besides the technique used for estimating the N amount are known to affect the estimates (Danso, 1995; Wani *et al.*, 1995; Shulz *et al.*, 1999a). The carry over effect on the following crop also depends upon the amount of legume biomass incorporated, N mineralisation and losses, as well as the non N effects of the legume crops (Schulz *et al.*, 1999b). Often, fallen leaves and the underground parts are not accounted for in such estimates. Pigeonpea and chickpea crops have shown a negative N balance when the above ground parts, often used as fodder for cattle or as fuel, were removed from the field (Wani *et al.*, 1995). N residual effect, fertilizer N replacement value (FRV) or N equivalent, have been used to measure and express the beneficial effects on the following cereal crop. Long term crop rotations involving chickpea and pigeonpea at the International Crops Research Institute for Semi Arid Tropics (ICRISAT) have clearly established the beneficial effects of these crops on sorghum yield, and mineral N content of the surface soil (Wani *et al.*, 1995). Similar results, and economic benefits have been shown for the northern cereal belt of Australia with chickpea and wheat (Herridge *et al.*, 1995b). Differences between grain legume species, and among the genotypes of common bean in percent N derived by fixation from the atmosphere, and total N fixed are reported (Danso, 1995). Fried (1995) has stressed the importance of enhancing the percent N derived by fixation from the atmosphere in the legume cultivars.

Table 1. Estimates of nitrogen fixed by the grain legumes crops and residual effect on cereal yield.

Grain legume crop	Estimates of N fixed (kg/ha)	Following cereal	Fertiliser N equivalent (kg/ha)
Black gram *(Vigna mungo)*	119-140	Sorghum	68
Chickpea *(Cicer arietinum)*	23-97	Maize	60-70
Common bean *(Phaseolus vulgaris)*	3-57	Pearl millet	40
Cow pea *(Vigna unguiculata)*	9-125	Pearl millet	60
Green gram *(Vigna radiata)*	50-66	Sorghum	68
		Wheat	68
		Pearl millet	30
Groundnut *(Arachis hypogaea)*	27-206	Pearl millet	60
		Wheat	28
		Maize	9-60
Lentil *(Lens culinaris)*	35-100	Pearl millet	40
		Maize	18-30
Peas *(Pisum sativum)*	46	Pearl millet	40
		Maize	20-32
Pigeonpea *(Cajanus cajan)*	4-200	Wheat	40
		Maize	20-67
Soybean *(Glycine max)*	49-450	Maize	7

Based on (Wani et al., 1995). Data compiled from experiments at different locations. Experiments with groundnut and pigeonpea were carried out at Patancheru, Andhra Pradesh, India.
Increase in rice dry matter yield of 16-21% and 12-20% after incorporating groundnut and soybean crop residues as compared to maize as pre-rice crop is reported by (Toomsan *et al.*, 1995).

Cultivar specific bacterial strains that enhance N fixation have been identified for several grain legume crops, and seed inoculation technologies have been developed. However, this technology has made a limited impact, especially in the tropical and sub-tropical regions (Bantilan and Johansen, 1995). Competition from the resident, native *Rhizobia* for the nodulation sites has restricted the role of the identified, superior inocluant strains as they fail to compete with the native strains for nodule occupancy. Even the high rates of *B. japonicum* inoculant used in soybean have been reported to result in less than 20% nodule occupancy (Lohrke *et al.*, 1996).

Nodulation mutants in grain legumes

Mutational analysis has been shown to be a powerful tool in genetic analysis of microbial functions. This is also true for the symbiotic interactions between the root nodulating bacteria and the legume host. The advance in the understanding of rhizobial genetics has been rapid where the symbiotic genes were located on large plasmids and on the chromosomes respectively. Freiberg et al. (1997) sequenced the 536kb plasmid NGR234 from a strain that is reported to nodulate over 100 legume genera. An international effort for sequencing of the complete genome of *Sinorhizobium meliloti* is underway (http://cmgm.stanford.edu/~mbarnett/genome.htm). Progress in molecular research of BNF has been relatively slow in the legume plants. However, 80 genes affecting symbiosis have been identified in different legumes (Tsyganov *et al.*, 1998). In pea more than 40 symbiotic genes have been identified, and the same number was anticipated in soybean (Rolfe and Gresshoff, 1988). As a result, the role of the host plant, and the importance of characterising nodulation mutants for understanding of BNF in legumes is now well recognised.

Genetic control of nodule number in pea was first reported by Gelin and Blixt (1964). Two nodulation resistant varieties of pea, one from Iran and the other from Afghanistan, were discovered by Lie (1971). This variety did not nodulate at 20°C after inoculation with a temperature sensitive strain of *R. leguminosarum* though normal nodulation was observed at 26°C. Another stock from Afghanistan showed a strain specific restriction, and did not nodulate after inoculation with the temperate strains. However, it showed normal nodulation after inoculation with the Middle-Eastern strain. Incompatibility between host plants of one region with the *Rhizobium* strains from a different region, and strain specificity became evident from the studies (Lie, 1984). These observations have led to the search of similar mutant types in the germplasm collections, and the use of chemical, and later physical, mutagens for inducing mutations affecting nodulation. The nodulation mutants were progeny tested, crossed with the parent and other lines to follow their inheritance, and appropriate gene symbols have been assigned. Mutants that nodulate in the presence of high nitrate concentration were first isolated by Carroll et al. (1985) after treatment of soybean seeds with ethyl methanesulfonate. The available information on nodulation mutants in different grain legumes is summarised in Tables 2-4 for pea, soybean and other grain legumes respectively. Pea and soybean are the most intensively investigated crops. It is pertinent to mention that such mutants have also been obtained in forage legumes, but they are not included in the present context of grain legumes.

Table 2. Symbiosis mutants in pea*

Parent/mutagen used mutant (genes)	Main characteristics	Reference
Mutants identified from germplasm collections from Afganistan, Iran and Turkey		
Sym-1	Temperature sensitive nodulation, resistance to some strains below 20° C	Lie, 1971, 1984
Sym-2	Nodulation resistance to a large number of European strains, dominance dependent upon Rhizobial strain, Ch.** 1. Arrest in infection thread, if *Rhizobium* strain does not produce Nod factors	Holl, 1975; Lie, 1984; Kneen and LaRue, 1984; Geurts *et al.*, 1997
sym-3	Ineffective nodulation, mr	Holl, 1975
sym-4	Nodulation resistance to specific Rhizobial strain	Lie, 1984
Prvus		Gelin and Blixt, 1964
nod1	Decreased nodulation, mr	
nod2	Decreased nodulation, mr	
Induced mutants		
Sparkle (EMS, gamma rays, fast neutrons)	Eleven mutants with few or no root nodules all recessive to wild type	Kneen *et al.*, 1994
E2*(sym-5)*	Nodulates at low temp, mr, Ch. 1,; inreased ethylene sensitivity, can be reverted to nodulation by blocking etylene accumulation	Kneen and LaRue, 1984 Fearn *et al.*, 1992; Heidstra *et* al., 1993
sym-6	Ineffective nodulation, mr	Caetano-Anolles and G Gresshoff, 1991
sym-7	Non-nodulation, mr, Ch. 3	
sym-8	Non-nodulation, Ch. 6, esseential for the induction of early nodulin genes PsENOD5 and PsENOD12A	Albrecht *et al.*, 1998
sym-9	Non-nodulation	Caetano-Anolles and Gresshoff, 1991
sym-10	Non-nodulation, mr, Ch. 10	
sym-11	Non-nodulation, mr,Ch. 7	
sym-12	Non-nonnodulation, mr	
sym-13	Ineffective nodulation, mr, Ch. 2	Kneen *et al.*, 1990

Parent/mutagen used mutant (genes)	Main characteristics	Reference
E135f	Normal number of ineffective white nodules, influences expression of leghaemoglobins and level of haem	Suganuma and LaRue, 1993 Suganuma *et al.*, 1995 Borisov *et al.*, 1992
E135N (*sym-14*)	Non -nodulation, mr	Kneen *et al.*, 1990
sym-15	Decreased nodulation, mr, Ch. 7	Caetano-Anolles and Gresshoff, 1991
sym-16	Decreased nodulation, mr, Ch. 5	
sym-17	Decreased nodulation, mr	
sym-18	Semidominant, nodulation resistance to some strains Ch. 1	
sym-19	non nodulation and resistance to Mycorrhizal infection, Ch. 1	
E107	Mutation at single recessive gene *brz*, has pleiotropic effects, along with low nodulation, partly restored by treatment with ethylene inhibitors	Guniel and LaRue, 1990, 1992
E132 (*sym21*)	Pleiotropic mutant with few nodules due to low rate of nodule meristem emergence	Markwei and LaRue, 1997
Frisson (EMS)		
	Seven nodulating, non-fixing mutants	Sagan *et al.*, 1993ab
P12(*sym-27*) P58(*sym-13*) P61(*sym-25*) P63(*sym-26*)	Nodule development normal followed by pre-mature senescence	Sagan *et al.*, 1994 Sagan and Duc, 1996
P64(*sym-28*) P88(*sym-29*)	Supernodulating, eight nitrate tolerant nodulation (Nod++Nts) mutants were identified, mr, genes	Sagan and Gresshoff, 1996
P6, P53, P54 P55, P56	Non nodulating, root hairs resistant to infection	Sagan *et al.*, 1994
P57	Low nodulating, root hair infection normal but early nodule development affected	
P59(*sym-23*)	Early arrest of nodule development	
P60(*sym-24*)		
P62	Developmentally defective root hairs	

Parent/mutagen used mutant (genes)	Main characteristics	Reference
myc-(1)	Mycorrhiza-resistant, VAM is blocked immediately after formation of apprasoria, mr, associated with nod, probably a common mechanism	Gianinazzi-Pearson *et al.*, 1991; Dumas-Gaudot *et al.*, 1994
myc-(2)	Intracellular hyphal development of VAM but no arbuscule formation	
Rondo (EMS)		Jacobsen, 1984; Jacobsen and Feenstra, 1984; Postma *et al.*, 1988
Nod3	Super nodulating, changed root and shoot morphology	
K5	Nodulation resistant, non-strain specific, infection thread formation	
K24	Nodulation resistant, only curling of root hair is observed	
Ramonskii 77 and K551 (EMS and NEU)		Sidorova and Uzhintseva, 1992; Sidorova *et al.*, 1995ab, 1997; Sidorova and Shumny, 1998
K301(nod4)	Supernodulating, nts nodules	
K287 (fix1)	Normal nodulation, white, ineffective nodules	
K1005m (sym)	Non-nodulating,	
	All three are mr, non allelic	
Sprint (EMS)		Borisov *et al.*, 1992, 1994 Sherrier *et al.*, 1997
Sprint 2 Nod-1 (*sym-8*)	Inability to form nodules and associations with vesicular-arbuscular mycorrhiza	
Sprint 2 Nod-2 (*sym-8)*		
Sprint 2 Fix⁻ (*sym-31*)	Impaired symbiosome development, Ch. 3	
RBT (*sym-13*, *sym-31*)	E135f x Sprint-2 Fix-, abnormal symbiosomes, suppresses manifestation of *sym-13* mutant allele.	
SGE (EMS)		Tsyganov *et al.*, 1998
SGEFix⁻1 (*sym40*)	Numerous white and some pink nodules, hypertrophied infection thread, premature degradation	
SGEFix⁻2 (*sym33*)	White nodules with dark pits, some pink nodules, characterised by "locked" infection thread.	

Parent/mutagen used mutant (genes)	Main characteristics	Reference
Finale (EMS, DES, ENU, NaN_3)		
	Thirty-six non nodulating and 24 mutants, with inefficient nodules	Engvild, 1987; Novak *et al.*, 1993ab, 1994, 1995, 1997
	Twenty one non-nodulating and 1 super nodulating	
RisfixC	Supernodulating, 5-6 times more nodules effective, nitrate resistant nodulation	
RisfixV	Hypernodulating, on low nitrate level, ineffective nodules, nodulation suppression by nitrate, traces of nitrogenase activity n nodules, mr	
RisfixO (*sym32*)	Ineffective nodulation, symbiotic tissue development arrested before formation of late symbiotic zone (LSZ)	
RisfixT	Ineffective nodulation, loss of LSZ after full) differentiation	
Double mutants	Obtained by crossing RisFixC x RisfixV	

**mr - monogenic recessive; **Ch - chromosome*

Table 3. Nodulation mutants in soybean*

Parent/mutagen used mutant (genes)	Main characteristics	Reference
Hyper-nodulating mutants		
Bragg (EMS)		
nts (**n**itrate **t**olerant **s**ymbiosis)		
	Super nodulating, NO3 tolerant. 15 independent nts mutants, all allelic and inherited as monogenic recessive, except *nts* 1116. *nts* 382, 183, 1007 and 1116 studied extensively. Absence of nodule autoregulation. Higher nodule number and biomass. *nts* 382 and 1007 are allelic. Not linked to nonnodulation mutants described later.	Caroll *et al.*, 1985, 1986, 1988; Delves *et al.*, 1986; Mathews *et al.*, 1990; Hansen and Akao, 1991;Lee *et al.*, 1991;

Parent/mutagen used mutant (genes)	Main characteristics	Reference
Williams (EMS and NMU)		
NOD1-3 2 - rj_7)	4 times more nodule number, nodules randomly distributed on both tap and lateral roots. Allelic to en650.	Gremaud and Harper, 1989; Wu and Harper, 1991; Ohmaya *et al.*, 1993; Pracht *et al*, 1993, 1994; Vuong *et al.*, 1996; Bacananamwo and Harper, 1997
NOD4(rj_7)	Single recessive locus, densely packed nodules, predominantly distributed along the entire length of the tap root. NOD1-3 and NOD4 are allelic.	
NOD2-4 (rj_8)	Independent allele, Crosses between NOD1-3 and NOD2-4 x NOD4, all F_1 were normal nodulation, in F_2 9 normal:7 hypernodulation was not observed. Authors speculate that rj_7 and rj_8 interact in an unknown manner.	
NOD3-7	Like other NOD mutants but no genetic analysis	
(All NOD mutants were selected under zero N conditions, but also show partial tolerance to NO_3. They are similar to the *nts* mutants obtained in cv. Bragg)		
Enrei (EMS)		
En6500 (rj_7)	Super nodulating, 6 fold increase larger number of nodules, with higher nodule dry wt. in presence of NO3, single recessive gene allelic to *nts*382	Akao and Kouchi, 1992; Kokubun and Akao, 1994; Takahashi *et al.*, 1995; Vuong *et al.*, 1996
Elgin 87 (EMS 0.04M for 12h)		
E300, E 420, E592	Super nodulating, allelic, recessive 6 - 10 x increase in nodule number	Buzzell *et al.*, 1990; Buttery and Buzzell, 1998
E391	Characters as above, but a second allele	
Sinpaldalkong 2 (EMS 30 mM)		
SS-2	Super-nodulation at early stages than *nts*1007 and *nts*1116, more nodule mass, higher C2H2 activity	Hong Suk *et al.*, 1997; Hong Suk and Suk Ha, 1998; Hong Suk *et al.*, 1998
Ineffective nodulation mutants		
Rj_2	Monogenic dominant	Caldwell, 1966; Caldwell *et al.*, 1966; Vest, 1970; Vest and Caldwell, 1972; Devine, 1984
Rj_3	Monogenic dominant	
Rj_4	Monogenic dominant	
Rj_5	Monogenic dominant	

Parent/mutagen used mutant (genes)	Main characteristics	Reference
Non-nodulating mutants		
rj_1	A non nodulating spontaneous mutant monogenic recessive, blockage at the stage of root hair curling	Williams and Lynch, 1954; Suganuma *et al.*, 1990
Williams (*N*-nitroso-*N*-methylurea)		
NN5	Does not form nodules in the field	Mathews *et al.*, 1989 (Pracht *et al.*, 1993)
(*rj_5, rj_6*)	Does not form nodules when inoculated with *Bradyrhizobium japonicum*, recessive, non allelic to *rj1*, allelic to *nod*139, a non nodulating mutant of Bragg	
Bragg (EMS, Gamma rays)		
*nod*49 (*rj_1*) *nod*772 (*rj_1*)	No nodulation after inoculation monogenic ressive inheritance in crosses with Bragg. Epistatically supress supernodulation in crosses with nts mutants. Shows few centres of sub-epidermal cell division after inoculation, but devoid of inocluation thread	Caroll *et al.*, 1986; Mathews *et al.*, 1989, 1992
*nod*139 (*rj_5,rj_6*)	Non allelic to *nod*49 and *nod*772 Devoid of infection thread and no subepidermal cell divisions after inoculation	

**nts mutants from Bragg, NOD mutants from Williams, and En6500 mutant from Enri revealed that all mutants, expressed similar levels of increased nodulation when grown under identical condition (Francisco and Harper, 1995). All mutants have been designated as hypernodulating by (Vuong et al., 1996)*

Table 4. Nodulation mutants in other grain legumes*

Parent/mutagen used/ mutant (genes)	Main characteristics	Reference
	Chickpea (*Cicer arietinum*)	
P502(ICC 640) Gamma rays		
PM 233 (*rn_1*) ***root nodule***	Nod-, mr	Davis *et al.*, 1985, 1986; Davis, 1988, 1991
PM 665 (*rn_2*)	Nod-, mr	
PM 679 (*rn_3*)	Nod-, mr, linked to simple leaves (*slv*)	
PM405 (*rn_4*)	Nod+, fix-, mr, small white nodules	
PM796(*rn_5*)	Nod+, fix-, mr, green nodules with narrow band of leghemoglobin	

Parent/mutagen used/ mutant (genes)	Main characteristics	Reference
ICC 640		
PM 233B	Nod-, blockage of *Rhizobium* infection process subsequent to root hair adsorption but prior to formation of infection thread and root cortical cell division	Mathew and Davis, 1990
The following are spontaneous mutants, identified in frequencies ranging between 120-490 per million plants.		
ICC 435		
ICC 435M (*rn*$_6$)	Nod-, mr, non allelic to PM 233, does not show N deficiency symptoms when grown in field	Rupela, 1992; Singh *et al.*, 1992
ICC4918		
ICC4918M	Nod-	
ICC4993		
ICC4993M	Nod-	
ICC5003		
ICC 5003M	Nod-	
Rabat		
Rabat NN (*rn*$_8$)	Nod-, mr	Singh and Rupela, 1998
Annigeri		
Annigeri NN	Nod-, mr	Singh and Rupela, 1998
P319-1NN	Nod-, mr, allelic to Annigeri NN, different gene for Nod- trait than that of RabbatNN and PM233	
Groundnut (*Arachis hypogaea*)		
All spontaneous mutants, or selected in the progenies of crosses between different normal nodulating genotypes.		
F$_3$ progenies of cross 487A-4-1-2 x PI 262090	Nod-, more than one gene	Gorbet and Burton, 1979
ICGL1 (Mutants selected in the F$_2$ generation of crosses involving ICGL1)		
ICGL1(GP-50, PI 544348)	Nod-, duplicate genes	Nigam *et al.*, 1980, 1982, 1991
ICGL2 (GP-51, PI 544349)	Nod-	
ICGL3 (GP-52, PI 544350)	Nod-	
ICGL4 (GP-53, PI, 544351)	Nod-	
ICGL5 (GP-54, PI 544352)	Nod-	

Parent/mutagen used/ mutant (genes)	Main characteristics	Reference
PI 109839	Trigenic	Dutta and Reddy, 1988
Georgia Non-Nod (Selected as a segregant in F_5 of a double cross)		
(GS-6, PI 595385)	Nod-, allelic to six diverse Nod-types from Georgia, Florida and ICRISAT, double recessive	Branch, 1997
	Cowpea (*Vigna ungiculata*)	
Iron & Clay varietal mix		
IC-1(*cpi*)	Fix-, ineffective nodulation, small size,no nitrogenase activity, mr, nodule anatomy similar to wild type, reduced size of the infected cells, number of bacteroids reduced, premature scenescence	Pemberton *et al.*, 1990; Purdom and Trese, 1995
JRW3-SmD	40-80% suppression of nodule formation depeding upon strain and time of inoculation	Singh and Ahmad, 1991
	Faba bean (*Vicia faba*)	
Feverole Ascott		
Feverole 13 (*sym-1*)	Fix-, ineffective nodulation, no acetylene reduction activity, effects either release of infective thread or differentiation of bacteria into bacteroids	Haser *et al.*, 1992
I 40		
I 40(*Sym-2*)	Nod-, monogenic dominant	Esser-Monning *et al.*, 1995
I 25		
I 25(*sym-3*)	Nod-, mr	
Ascott (EMS)		Duc, 1995
f13(*sym-1*) f73(*sym-2*) f245(*sym-3*)	Nod+/Fix-, mr	
f48(*sym-4*)	Nod-, mr	
f32(*sym-5*)	Nod+ +, nts, 3-5 times more nodules than control, in absence of combined nitrogen and nitrate tolerant nodulation, mr	
	Common bean (*Phaseolus vulgaris*)	
RIZ30 (EMS)		
NOD238(*sym-2*)	Fix -, ineffective nodulation, small nodules	Pedalino and Kipe-Nolt, 1993; Pedalino *et al.*, 1993
NOD125(*sym-1*) Nod-		Park and Buttery, 1988, 1997

Parent/mutagen used/ mutant (genes)	Main characteristics	Reference
RIZ36 (EMS)		
NOD109	Fix -, late formation of pink nodules,low AR activity, poor pod fertility	Pedalino and Kipe-Nolt, 1993; Pedalino *et al.*, 1993
R69(nie)	Ineffective nodulation, mr epistatic to nts, non allelic to sym-2	Park and Buttery, 1997; Pedalino *et al.*, 1992
R99(nnd-2)	Non nodulation, mr, epistatic to nts and nie	
OAC Rico (EMS)		
R699(*nie*)	nod+/fix-, Myc-small, white nodules without central infected zone, IN, mr	Park and Buttery, 1988, 1989a, 1994
R99(*nnd-2*)	nod-, NN, Myc+	Shritliffe and Vessey, 1996; Shritliffe *et al.*, 1996
R32 (GP-88, PI 536541) (*nts*)	Nitrate tolerant super nodulation (NTSN), mr, delayed maturity, and slightly lower yield	
Swan Valley (EMS)		
SV145	NTSN, mr	Park and Buttery, 1989b, 1994
	Pigeonpea (*Cajanus cajan)*	
Non-nodulating with native root nodulating bacteria (RNB) Low nodulating at low N High nodulating at low N		Wani *et al.*, 1995

*AR: acetylene reduction; NN: non-nodulating; NTSN - nitrate tolerant supernodulation; mr: monogenic recessive

The nodulation mutants are broadly classified into four classes (Sagan *et al.*, 1994):

nod-	no nodules
nod+/-	few or no nodules
fix-	ineffective nodulation
nod++	supernodulation or hypernodulation
nts(**n**itrate **t**olerant **s**ymbiosis)	nitrate tolerant nodulation.

In addition, other mutants showing resistance to mycorrhiza, and deficient in nitrate reductase that have been identified are included in Tables 2 - 4:

myc-	resistant to mycorrhiza
nar-	nitrate reductase deficient
etr	ethylene resistance (insensitivity)

An ethylene insensitive mutant of *Medicago trunculata* is hyperinfected with *Rhizobium* (Penmesta and Cook, 1997). In pea, nodulation mutants can be reverted to normal nodulation by ethylene inhibitors (Fearn *et al.*, 1992; Guniel and LaRue, 1992). Ethylene-insensitive mutants that show altered pathogen susceptibility have been isolated in soybean (Hoffman *et al.*, 1999). However, there were no differences in the nodule number on a variety of soybean mutants showing strong, intermediate and weak ethylene insensitivity (Schmidt *et al.*, 1999). Investigators who have recorded nodulation in field experiments with native soil microbes have used a more descriptive classification (Wani *et al.*, 1995) as follows:

Non nodulating with native root nodulating bacteria (RNB)
Non nodulating with a specific strain, low nodulating with native RNB
Low nodulating at low N
High nodulating at low N but low nodulating at high N
High nodulating at high N

Inheritance of nodulation mutants

Spontaneous as well as induced nodulation mutants are mostly monogenic recessive (Tables 2-4), though involvement of two and more genes (Dutta and Reddy, 1988), and some dominant (Caldwell, 1966a; Devine, 1984; Esser-Monning *et al.*, 1995) mutations are also reported. Many of the induced hypernodulating mutants in different cultivars of soybean have been found to be allelic (Devine and Kuykendall, 1996). Nonnodulating soybean mutants (nod49 and nod772) were allelic but not nod139. Hypernodulating nts382 was not allelic to any of these nonodulating mutants, and nonnodulation was found to be epistatic over hypernodulation (Mathews *et al.*, 1990). No spontaneous hypernodulating mutant has been reported. Most hypernodulating mutants were obtained earlier after treatment with chemical mutagens (Caetano-Anolles and Gresshoff, 1991). However, fast neutron induced hypernodulating mutants in *Glycine soja* have been reported recently (Gresshoff *et al.*, 1999).

Natural and induced mutations isolated in different grain legume crops show no, decreased, ineffective, hyper and nitrate tolerant nodulation. The observed outcome results from the interactions of two genetic systems and is bacterial strain dependent. The non-nodulation mutants restrict nodule formation after inoculating either with a specific or with many different bacterial strains. In nod-, and fix- mutants the nodulation process is blocked at any one of the stages of nodule development described earlier. Some of the nod- mutants show no differences in early signal exchange or other pre-infection events. Non-nodulation is controlled by the root genotype (Cho and Harper, 1991; Mathews *et al.*, 1992). The root exudates of nod- mutants show similar *nod* gene inducing activity. Others show few subepidermal cell divisions, but infection threads are not formed. Beside mutants with developmentally defective root hair development, mutants with a low rate of nodule meristem emergence, and pre-mature senescence of nodules have been isolated. Many nodulation mutants have altered root characteristics, and many plant hormones known to influence root development also affect nodulation.

Use of nodulation mutants

The nodulation mutations have been used to understand the role of specific genes in the development and functioning of the root nodules. Suggestions have been made to make use of the restrictive nodulation mutants, and to combine several genes to eliminate competition with the native soil rhizobia (Lohrke *et al.*, 1996). Low competitive ability of the selected or genetically tailored strains in the field is widely recognized as the major constraint in enhancing BNF by the grain legume crops (Fried, 1995). A crop cultivar having gene(s) that restrict nodulation from the native soil rhizobia will be nodulated only by the identified inoculated strain.

Inhibition of nodulation and N fixation in presence of soil nitrates is the second main constraint to the utilization of BNF. Availability of nitrate inhibits early steps of nodulation as well as N fixation (Streeter, 1988; Caroll and Mathews, 1990). Hence, hypernodulating and nts mutants are of great interest. Pierce and and Bauer (1983) and Caetano-Anolles and Greshoff (1991) have discussed the evidence that suggests feedback control of nodule development after the formation of a critical number of nodules. This is referred to as autoregulation and was first observed by Nutman (1952) in red clover; surgical removal of functional nodules restored formation of new nodules. The *nts* mutants in soybean are reported to diminish autoregulation. Hypernodulation was observed when shoots of hypernodulating mutants of soybean and common bean were grafted on normal and non nodulating genotypes, indicating that hypernodulation is primarily controlled by the shoot (Delves *et al.*, 1986; Day *et al.*, 1989; Cho and Harper, 1991). Grafting of the shoots of hyper nodulating soybean mutant on normal nodulating soybean, mungbean and hyacinth bean, has shown that a common, shoot generated signal that is translocated through the graft, induces hypernodulation in all the species (Harper *et al.*, 1997). The substance involved in autoregulation is reported to be different in soybean and common bean (Hamaguchi *et al.*, 1993). Hypernnodulation mutants have been used to find answers to many questions related to nodulation (Caetano-Anolles and Gresshoff, 1991; Gresshoff *et al.*, 1999). The *nts-1* gene has been mapped on the linkage group H of the USDA soybean map 0.3 to 0.4 cM from pUTG132a marker (Gresshoff *et al.*, 1999) and efforts are on map-based cloning of the gene. No host nodulation genes have been cloned nor their products yet identified in grain legumes. Non-nodulating mutant blocking the infection thread and nodule primordia formation, designated as *nin* (nodule inception), has been recently reported in forage legume *Lotus japonicus* (Schauser et al. 1999). The mutant was isolated after transposon tagging. Their results suggest that the NIN protein acts as a transcription regulator of genes required for nodule development.

Field assessment of the hypernodulating mutants

The initial performance of the hypernodulating soybean mutants NOD1-3, NOD2-4 and NOD3-7 (Wu and Harper, 1991) and NOD4 and NOD1-3 (Pracht *et al.*, 1994) in field experiments were disappointing. Yield of normal nodulating parent 'Williams' and other cultivars were significantly higher than the hypernodulating mutants. Similarly the yield of the nts mutants, though producing five times the number of nodules and eight times the nodule weight in comparison to the check cv. 'Centaur,' did not show consistent, increase in productivity. Grain yields ranged between 25% less to equal to those of the normal nodulating cultivars. However, the carry over effect on the following crop of oats and barley was significantly higher after the hypernodulating genotypes, in comparison to the normal-nodulating cultivars (Song *et al.*, 1995). Subsequently a hypernodulating (nts) soybean cultivar 'Nitrobean-60' has been released in Australia (Devine and Kuykendall, 1996; Anon., 1997b). Low productivity of the hypernodulating types have been attributed to physiological growth limitation due to excessive nodulation. Other deleterious mutations, unrelated to the hypernodulation could also be responsible for low productivity (Pracht *et al.*, 1994). Reduction in yield of the selected mutants, relative to the parent, has often been observed in induced mutation experiments. Usually it is possible to recover the yield losses after backcrossing the mutant to the parent and other cultivars. The same may be true also for the hypernodulating mutants.

Energetics

A detailed discussion on the energetics of nitrogen fixation would be out of place in the present context. However, it is important to remember that the N fixed by the grain legumes, which appears to be free of cost to the farmer, has high cost in terms of the carbon assimilates provided by the crop for using atmospheric N. Nitrogenase requires 36 ATP molecules for every N_2 molecule reduced. Photosynthates are used for the growth and maintenance of the nodules, nitrogenase/hydrogenase, ammonia assimilation and transport of N. Non-nodulated plants growing on soil nitrates also use C assimilates for nitrate uptake, reduction, ammonia assimilation and transport. Theoretical estimates for the cost of N fixation by the legumes range between 2.6 - 6.1 gC/gN, in comparison to 0.8 - 2.4 gC/gN for N assimilation from soil nitrate (Atkins, 1984). Kennedy (1997) has estimated the need of 6.5 gC/gN. Atkins (1984) has inferred "that while the most efficient symbioses would be energetically similar to the most expansive example of nitrate reduction, many would be likely to require a greater allocation of C". Experimental comparisons of C and N economy in nodulated and nonnodulated grown plants and between different grain legume species have been made (Atkins *et al.*, 1979; Pate *et al.*, 1979; Atkins *et al.*, 1980) which show variation in the amount of C per gN between grain legume species, and in the same species at different stages of plant growth.

Further, in grain legumes with determinate growth habit and monocarpic senescence, after anthesis, competition between root nodules and developing seeds for C assimilates sets in (Sinclair and de Wit, 1975). The nodules are deprived of C assimilates, and N mobilization from foliage to developing seeds accelerates leaf senescence. Other studies have suggested that 19.5% of the gross photosynthates are required for N fixation in soybean (Sheehy, 1999) and early withdrawal of foliage N may not always be essential for seed development. Lines showing delayed leaf senescence have been selected (Abu-Shakra *et al.*, 1978). Physiological approaches for mobilization of stored N in grain legume plants have been suggested (Shantharam and Matto, 1997). With the range of mutants available, or that can be induced, it should be possible to develop genotypes, within the energetic constraints, where photosynthates are not limiting for N fixation. This is not going to be simple but the expected benefits at the end are worth sustained efforts.

Perspective for the future

Considering the overall importance of N fixed by the grain legume crops and their role in maintaining the long-term soil health and N fertility (Vance, 1997), grain legume breeders should pay greater attention to the proportion of N derived from fixation. This is all the more important for evolving more productive, sustainable cropping systems in areas where fertilizer availability is limited, or in organic farming. Yield of the following non-legume crop obtained without any fertiliser N application can be used as the gross measure of N contributed by the selected genotypes to the soil. N fixation, like other complex characters, such as yield, has several components - early nodulation, optimization of nodule number and weight, occupancy of nodulation sites by the most efficient bacterial strain, and longer duration of N fixation activity (delayed senescence of nodules). These along with the ^{15}N methods available to assess the percent N derived by fixation from the atmosphere can be used as selection parameters. Nonnodulation and hypernodulation mutants, along with the new biotechnology tools and techniques, now available can be used for the selection of the best host and bacterial combinations.

References

Anon., 1984. Breeding Legumes for Enhanced Symbiotic Nitrogen Fixation. Martinus Nijhoff/Dr W. Junk Publishers, Dordrecht

Anon., 1994. Nuclear Techniques in Soil Plant Studies for Sustainable Agriculture and Environment Preservation. IAEA, Vienna

Anon., 1997. Biological Fixation of Nitrogen for Ecology and Sustainable Agriculture. Springer Verlag, Berlin

Anon., 1997. List of new mutant cultivars; *Glycine max* L. (soybean). MBNL. **43**: 44-45

Anon., 1997. Sustainable Agriculture for the Tropics: the Role of Biological Nitrogen Fixation. Elsevier Science Ltd.

Anon., 1999 Management of Biological Nitrogen Fixation for the Development of More Productive and Sustainable Agricultural Systems. Kluwer Academic Publishers

Abu-Shakra, S.S., D.A. Phillips, and R.C. Huffaker, 1978. Nitrogen fixation and delayed leaf senescence in soybeans. Science. **199**: 973-975

Akao, S. and H. Kouchi, 1992. A supernodulating mutant isolated from soybean cultivar Enri. Soil Sci.and Plant Nutr. **38**: 183-187

Albrecht, C., R. Geurts, and F. Lapeyrie, 1998. Endomycorrhizae and rhizobial Nod factors both require *sym8* to induce the expression of the early noduling genes *PsENODS* and *PsENOD12A*. The Plant J. **15**: 695-614

Ali, M., 1999. Evaluation of green manure technology in tropical lowland rice system. Field.Crops Res. **61**: 61-78

Atkins, C.A., 1984. Efficiences and inefficiencies in the legume/*Rhizobium* symbiosis - a review. Plant and Soil. **82**: 273-284

Atkins, C.A., J.S. Pate, G.J. Griffiths, and S. T. White, 1980. Economy of carbon and nitrogen in nodulated and non-nodulated $N0_3$ grown cowpea (*Vigna unguiculata* (L.) Walp.). Plant Physiol. **66**: 978-983

Atkins, C.A., J.S. Pate, and D.B. Lyzell, 1979. Assimilation and transport of nitrogen in non-nodulated (NO_3 grown) *Lupinus albus* L. Plant Physiol. **64**: 1078-1082

Bacananamwo, M. and J.E. Harper, 1997. Response of a hypernodulating soybean mutant to increased photosynthate supply. Plant Sci. **124**: 119-129

Bantilan, M.C.S. and C. Johansen, 1995. Research evaluation and impact of biological nitrogen fixation. Plant and Soil. **174**: 279-286

Bhatia, C.R., K. Nichterlein, L. van Zanten, and M. Maluszynski, 1999. Grain legume cultivars developed using induced mutations. Mut.Breed.Rev. **11**

Boddey, R.M., J.C. de Moraes Sa, J. Bruno, R. Alves, and S. Urquiaga, 1997. The contribution of biological nitrogen fixation for sustainable agriculture systems in the tropics. Soil Biol. Biochem. **29**: 787-799

Borisov, A.Y., E.V. Morzhina, O.A. Kulikova, S.A. Tchetkova, V.K. Lebsky, and I. A. Tikhonovich, 1992. New symbiotic mutants of pea (*Pisum sativum* L.) affecting either nodule initiation or symbiosome development. Symbiosis (Rehovot). **14**: 297-313

Borisov, A.Y., S.M. Rozov, V.E. Tsyganov, O.A. Kulikova, A.N. Kolycheva, L.M. Yakobi, A.O. Ovtsyna, and I.A. Tikhonovich, 1994. Identification of symbiosis genes in pea (*Pisum sativum* L.) by means of induced mutagenesis. Genetika. **30**: 1484-1494

Branch, W.D., 1997. Registration of Georgia Non-Nod peanut genetics stock. Crop Sci. **37**: 1412

Brevin, N.J., 1991. Development of legume root nodule. Annu. Rev. Cell Biol. **7**: 191-226

Buttery, B.R. and R.I. Buzzell, 1998. Inheritance of autoregulation mutants in Elgin 87 soybean resulting in supernodulation phenotypes. Soybean Genet. Newsl. **25**: 25-26

Buzzell, R.I., B.R. Buttery, and G. Ablett, 1990. Supernodulation mutants in Elgin 87 soybean. In: Nitrogen Fixation: Achievements and Objectives. Gresshoff, P.M., L.E. Roth, G. Stacey and W.E. Newton (Eds.) Chapman and Hill, New York. pp.726.

Caetano-Anolles, G. and P.M. Gresshoff, 1991. Plant genetic control of nodulation. Annu. Rev. Microbiol. **45**: 345-382

Caetano-Anolles, G. and P.M. Gresshoff, 1996. Advances in positional cloning of nodulation genes in soybean. Plant and Soil. **186**: 1-7

Caldwell, B.E., 1966. Inheritance of a strain specific ineffective nodulation in soybean. Crop Sci. **6**: 427-428

Caldwell, B.E., K. Hinson, and K. Johnson, 1966. A strain-specific ineffective nodulation reaction in the soybean (*Glycine max* L. Merrill). Crop Sci. **6**: 495-496

Caroll, B.J., P.M. Gresshoff, and A.C. Delves, 1988. Inheritance of supernodulation in soybean and estimation of genetically effective cell number. Theor. Appl. Genet. **76**: 54-58

Caroll, B.J. and A. Mathews, 1990. Nitrate inhibition of nodulation in legumes. In: Molecular Biology of Symbiotic Nitrogen Fixation. Gresshoff, P. (Ed.) CRC Press, Boca Raton. pp.159-180

Caroll, B.J., D.L. McNeil, and P.M. Gresshoff, 1985. Isolation and properties of soybean [*Glycine max* (L.) Merr.] mutants that nodulate in the presence of high nitrate concentrations. Proc. Natl. Acad. Sci. USA. **82**: 4162-4166

Caroll, B.J., D.L. McNeil, and P. M. Gresshoff, 1986. Mutagenesis of soybean (*Glycine max* (L.) Merr.) and isolation of non-nodulating mutants. Plant Sci. **47**: 109-114

Cho, M.J. and J.E. Harper, 1991. Root isoflavonoid response to grafting between wild-type and nodulation mutant soybean plants. Plant Physiol. **96**: 1277-1282

Danso, S.K.A., 1995. Sustainable agriculture: the role of biological nitrogen fixing plants. In: Nuclear Techniques in Soil Plant Studies for Sustainable Agriculture and Environment Preservation. IAEA, Vienna. pp.205-224

Davis, T.M., 1988. Two genes that confer ineffective nodulation in chickpea (*Cicer arietinum* L.). J. Hered. **79**: 476-478

Davis, T.M., 1991. Linkage relationships of genes for leaf morphology flower color and root nodulation in chickpea. Euphytica. **54**: 117-124

Davis, T.M., K.W. Foster, and D.A. Phillips, 1985. Nonnodulation mutants in chickpea. Crop Sci. **25**(345): 348

Davis, T.M., K.W. Foster, and D.A. Phillips, 1986. Inheritance and expression of three genes controlling root nodule formation in chickpea. Crop Sci. **26**: 719-723

Day, D.A., B.J. Carroll, A.C. Delves, and P.M. Gresshoff, 1989. Relationship between autoregulation and nitrate inhibition of nodulation in soybeans. Physiol. Plant. **75**: 37-42

Delves, A.C., A. Mathew, D.A. Day, A.S. Garter, B.J. Caroll, and P.M. Gresshoff, 1986. Regulation of soybean-*Rhizobium* nodule symbiosis by shoot and root factors. Plant Physiol. **82**: 588-590

Denaire, J., F. Debelle, and C. Rosenberg, 1992. Signalling and host range variation in nodulation. Ann. Rev. Microbiol. **46**: 497-531

Devine, T.E., 1984. Inheritance of soybean nodulation response with a fast growing strain of *Rhizobium*. J. Hered. **75**: 359-361

Devine, T.E. and L.D. Kuykendall, 1996. Host genetic control of symbiosis in soybean (*Glycine max* L.). Plant and Soil. **186**: 173-187

Djordjevic, M.A., D.W. Gabriel, and B.G. Rolfe, 1987. *Rhizobium*, the refined parasite of legumes. Annu. Rev. Phytopathol. **25**: 145-168

Dobereiner, J., 1997. Biological nitrogen fixation in the tropics: social and economic contributions. Soil Biol. Biochem. **29**: 771-774

Duc, G., 1995. Mutagenesis of faba bean (*Vicia faba* L.) and the identification of five different genes controlling no nodulation, ineffective nodulation or super nodulation. Euphytica. **83**: 147-152

Dumas-Gaudot, E., A. Asselin, V. Gianinazzi-Pearson, A. Gollotte, and S. Gianinazzi, 1994. Chitinase isoforms in the roots of various pea genotypes infected with arbuscular mycorrhizal fungi. Plant Sci. **99**: 27-37

Dutta, M. and L.J. Reddy, 1988. Further studies on genetics of nonnodulation in peanut. Crop Sci. **28**: 60-62

Engvild, K.C., 1987. Nodulation and nitrogen fixation mutants of pea, *Pisum sativum*. Theor. Appl. Genet. **74**: 711-713

Esser-Monning, K., P. Roskothen, and G. Röbbelen, 1995. Two host genes in *Vicia faba* for nodulation deficiency with strain specificity for *Rhizobium leguminosarum*. Plant Breed. **114**: 363-365

Fearn, J.C., F.C. Guinel, and T.A. LaRue, 1992. Ethylene antagonists increase nodulation on *sym5* pea mutants. In: Proc.14th Intern. Conf. Plant Growth Substances, Amsterdam. Kluwer Academic Publishers, Dordrecht. pp.530-533

Francisco, P.B. Jr. and J.E. Harper, 1995. Translocatable leaf signal autoregulates soybean nodulation. Plant Sci. **107**: 297-311

Freiberg, C., R. Fellay, A Bairoch, W. J Broughton, A. Rosenthal, and X. Perret, 1997. Molecular basis of symbiosis between *Rhizobium* and legumes. Nature. **387**: 394-401

Fried, M., 1995. Biological nitrogen fixation: present and future. In: Nuclear Techniques in Soil Plant Studies for Sustainable Agriculture and Environment Preservation. IAEA, Vienna. pp.199-204

Gelin, O. and S. Blixt, 1964. Root nodulation in peas. Agri. Hortique Genetica. **22**: 149-159

Geurts, R., R. Heidstra, A.E. Hadri, S.R. Downie, H. Franssen, A. Van Kammen, and T. Bisseling, 1997. *Sym2* of pea is involved in a nodulation factor-perception mechanism that controls the infection process in epidermis. Plant Physiol. **115**: 351-359

Gianinazzi-Pearson, V., S. Gianinazzi, J.P. Guillemin, A. Trouvelot, and G. Duc, 1991. Genetic and cellular analysis of resistance to vesicular-arbuscular (VA) mycorrhizal fungi in pea mutants. In: Advances in Molecular Genetics of Plant-Microbe Interactions. Vol. I. Hennecke, H. and D.P.S. Verma (Eds.) Kluwer Academic Publishers, Dordrecht. pp.336-342

Gorbet, D.W. and J.C. Burton, 1979. A non nodulating peanut. Crop Sci. **19**: 727-728

Gremaud, M.F. and J.E. Harper, 1989. Selection and initial characterization of partially nitrate tolerant nodulation mutants of soybean. Plant Physiol. **89**: 169-173

Gresshoff, P.M., 1993. Molecular genetic analysis of nodulation genes in soybean. Plant Breed. Rev. **11**: 275-318

Gresshoff, P.M., J. Stiller, Q. Jiang, A. Hussain, D. Lohar, C.A. Vasile, I.L. Anderssen, C. Zaharia, M. Appel, K. Harada, I. Searle, M. Bhattacharrya, K. Meksem, D. Lightfoot, M. Spencer, B. Aroll, and A. Men, 1999. Functional genomics and genetic analysis of nodulation and lateral root development of soybean and *Lotus japonicus*. Symbiosis (Rehovot) (in press)

Guniel, F.C. and T.A. LaRue, 1990. Formation of wall appositions in leaves and lateral roots of an iron-accumulating pea mutant. Can.J.Bot. **68**: 1340-1348

Guniel, F.C. and T.A. LaRue, 1992. Ethylene inhibitors partly restore nodulation of pea mutant E107(brz). Plant Physiol. **99**: 515-518

Hamaguchi, H., M. Kokubun, T. Yoneyama, A.P. Hansen, and S. Akao, 1993. Control of supernodulation in intergeneric grafts of soybean and common bean. Crop Sci. **33**: 794-797

Hansen, A.P. and S. Akao, 1991. Early nodule development of soybean cv. Bragg and some nodulation mutants in response to different levels of inoculation. J. Plant Physiol. **138**: 501-506

Harper, J.E., K.A. Corrigan, A.C. Barbera, and M.H. Abd-Alla, 1997. Hypernodulation of soybean, mungbean and hyacinth bean is controlled by a common shoot signal. Crop Sci. **37**: 1242-1246

Haser, A., D.L. Robinson, G. Duc, and C. P. Vance, 1992. A mutation in *Vicia faba* results in ineffective nodules with impaired bacteroid differentiation and reduced synthesis of late nodulins. J. Exp. Bot. **43**: 1397-1407

Heidstra, R., A. Kozik, T. Bisseling, and T. A. Lie, 1993. Pea Nod-mutants and Nod factor perception. In: New Horizons in Nitrogen Fixation. Palacios, R., J. Mora and W.E. Newton (Eds.) Kluwer Academic Publisher, Dordrecht. pp.267-268

Herridge, D.F. and S.K.A. Danso, 1995. Enhancing crop N_2 fixation through selection and breeding. Plant and Soil. **174**: 51-82

Herridge, D.F., H. Marcellos, W. Felton, G.D. Schwenke, M. Aslam, S. Ali, Z. Shah, H. Shah, S. Maskey, S. Bhattari, M. Peoples, and G.L. Turner, 1995. Management of legume N_2 fixation in cereal systems. A research programme for the rain fed areas of Pakistan, Nepal and Australia. In: Nuclear Techniques in Soil Plant Studies for Sustainable Agriculture and Environment Preservation. IAEA, Vienna. pp.237-250

Hoffman, T., J.S. Schmidt, and X. Zheng, 1999. Isolation of ethylene-insensitive soybean mutants that are altered in pathogen susceptibility and gene-for-gene disease resistance. Plant Physiol. **119**: 935-949

Holl, F.B., 1975. Host plant control of the inheritance of dinitrogen fixation in *Pisum-Rhizobium* symbiosis. Euphytica. **24**: 767-770

Hong Suk, L. and L. Suk Ha, 1998. Introduction, development, and characterization of supernodulating soybean mutant - nitrate inhibition of nodulation and nitrogen fixation in supernodulating soybean mutant. Korean J. Crop Sci. **43**: 23-27

Hong Suk, L., K Yong Wook, and P. Eui Ho, 1998. Introduction, development, and characterization of supernodulating soybean mutant - shoot factor regulation of nodule development in supernodulating soybean mutant. Korean J. Crop Sci. **43**: 28-31

Hong Suk, L., C. Young Am, P. Eui Ho, K. Yong Wook, I. Kwang, and L. Suk Ha, 1997. Introduction, development, and characterization of supernodulating soybean mutant. I. Mutagenesis of soybean and selection of supernodulating soybean mutant. Korean J. Crop Sci. **42**: 247-253

Jacobsen, E., 1984. Modification of symbiotic interaction *Pisum sativum* and *Rhizobium leguminosarum* by induced mutation. Plant and Soil. **82**: 427-438

Jacobsen, E. and W.J. Feenstra, 1984. A new pea mutant with efficient nodulation in the presence of nitrate. Plant Sci Lett. **33**: 337-334

Kennedy, I.R., 1997. BNF: an energy costly process? In: BNF: The Global Challenge and Future Needs. A position paper discussed at the Rockefeller Foundation Bellagio Conference Centre, Lake Como, Italy, April 8-12.

Kneen, B.E. and T.A. LaRue, 1984. Pea (*Pisum sativum* L.) with strain specificity for *Rhizobium leguminosarum*. Heredity. **52**: 383-389

Kneen, B.E., T.A. LaRue, A.M. Hirsch, C.A. Smith, and N.F. Weeden, 1990. *sym13* - A gene conditioning ineffective nodulation in *Pisum sativum*. Plant Physiol. **94**: 899-905

Kneen, B.E., N.F. Weeden, and T.A. LaRue, 1994. Non-nodulating mutants of *Pisum sativum* (L.) cv. Sparkle. J. Hered. **85**: 129-133

Kokubun, M. and S. Akao, 1994. Inheritance of supernodulation in soybean mutant En6500. Soil Sci. and Plant Nutr. **40**: 715-718

Lee, S.H., D.A. Ashley, and H. R Boerma, 1991. Regulation of nodule development in supernodulating mutants and wild type soybean. Crop Sci. **31**: 688-693

Lie, T.A., 1971. Temperature dependent root-nodule formation in pea cv. Iran. Plant Soil. **34**: 751-752

Lie, T.A., 1984. Host genes in *Pisum sativum* conferring resistance to European *Rhizobium leguminosarum*. Plant Soil. **82**: 415-425

Lohrke, S.M., J.H. Orf, and M.J. Sadowsky, 1996. Inheritance of host controlled restriction of nodulation by *Bradyrhizobium japonicum* strain USDA 110. Crop Sci. **36**: 1271-1276

Long, S.R., 1989. *Rhizobium* genetics. Annu. Rev. Genet. **23**: 483-506

Markwei, C.M. and T.A. LaRue, 1997. Phenotypic characterization of *sym21*, a gene conditioning shoot-controlled inhibition of nodulation in *Pisum sativum* cv. Sparkle. Physiol. Plant. **100**: 927-932

Mathew, L.J. and T.M. Davis, 1990. Anatomical comparison of wild type and nonnodulating mutant chickpea (*Cicer arietinum*). Can. J. Bot. **68**: 1201-1207

Mathews, A., B.J. Carroll, and P.M. Gresshoff, 1989. Development of *Bradyrhizobium* infections in supernodulating and non-nodulating mutants of soybean (*Glycine max* (L.) Merrill). Protoplasma. **150**: 40-47

Mathews, A., B.J. Carroll, and P.M. Gresshoff, 1990. The genetic interaction between non-nodulation and supernodulation in soybean: an example of developmental epistasis. Theor. Appl. Genet. **79**: 125-130

Mathews, A., B.J. Carroll, and P.M. Gresshoff, 1992. Studies on root control of non-nodulation and plant growth of non-nodulating mutants and a supernodulating mutant of soybean (*Glycine max* (L.) Merr.). Plant Sci. **83**: 35-43

Micke, A., 1988. Genetic improvement of grain legumes using induced mutations. An overview. In: Improvement of Grain Legume Production Using Induced Mutations. IAEA, Vienna. pp.1-51

Nap, J.P. and T. Bisseling, 1990. Developmental biology of a plant-prokaryote symbiosis: the legume nodule. Science. **250**: 948-954

Nigam, S.N., V. Arunachalam, R.W. Gibbons, A. Bandhopadyay, and P.C.T. Nambiar, 1980. Genetics of non-nodulation in groundnut (*Arachis hypogaea* L.). Oleagineux. **35**: 453-455

Nigam, S.N., S. L. Dwivedi, P.C.T. Nambiar, and R.W. Gibbons, 1991. Registration of five nonnodulating peanut germplasm lines. Crop Sci. **31**: 855-856

Nigam, S.N., P.C.T. Nambiar, S.L. Dwivedi, and P. Dart, 1982. Genetics of non nodulation in groundnut (*Arachis hypogaea* L.): studies with single and mixed *Rhizobium* strains. Euphytica. **31**: 691-693

Novak, K., K. Pesina, J. Nebesarova, V. Skrdleta, L. Lisa, and V. Nasinec, 1995. Symbiotic tissue degradation pattern in the ineffective nodules of three nodulation mutants of pea (*Pisum sativum* L.). Ann. Bot. **76**: 303-313

Novak, K., V. Skrdleta, M. Kropacova, L. Lisa, and M. Nemcova, 1997. Interaction of two genes controlling symbiotic nodule number in pea (*Pisum sativum* L.). Symbiosis (Rehovot). **23**: 43-62

Novak, K., V. Skrdleta, and L. Lisa, 1994. Plant growth, photosynthetic pigments, and nitrate assimilation in symbiotic mutants of garden pea. J. Plant Nutr. **17**: 1265-1273

Novak, K., V. Skrdleta, M. Nemcova, and L. Lisa, 1993a. Behaviour of pea nodulation mutants as affected by increasing nitrate level. Symbiosis (Rehovot). **15**: 195-206

Novak, K., V. Skrdleta, M Nemcova, and L. Lisa, 1993b. Symbiotic traits, growth, and classification of pea nodulation mutants. Rostlinna Vyroba. **39**: 157-170

Nutman, P.S., 1952. Studies on the physiology of nodule formation. III. Experiments on excision of root tips and nodules. Ann. Bot. **16**: 80-102

Ohmaya, T., J.C. Nicholas, and J. E. Harper, 1993. Assimilation of $^{15}N_2$ and $^{15}NO_3$ by partially nitrate tolerant nodulation mutants of soybean. J. Exp. Bot. **44**: 1739-1747

Park, S.J. and B.R. Buttery, 1988. Nodulation mutants of white bean (*Phaseolus vulgaris* L.) induced by ethyl-methane sulphonate. Can. J. Plant Sci. **68**: 199-202

Park, S.J. and B.R. Buttery, 1989a. Identification and characterization of common bean (*Phaseolus vulgaris* L.) lines well nodulated in the presence of high nitrate. Plant and Soil. **119**: 237-244

Park, S.J. and B.R. Buttery, 1989b. Inheritance of nitrate tolerant supernodulation in EMS-induced mutants of common bean (*Phaseolus vulgaris* L.). J. Hered. **80**: 486-488

Park, S.J. and B.R. Buttery, 1994. Inheritance of non-nodulation and ineffective nodulation mutants in common bean. J. Hered. **85**: 1-3

Park, S.J. and B.R. Buttery, 1997. Complementation of nodulation genes of various mutants in common bean (*Phaseolus vulgaris* L.). J. Hered. **88**(6): 543-545

Pate, J.S., D.B. Layzell, and C.A. Atkins, 1979. Economy of C and N in a nodulated and non-nodulated (NO_3 - grown) legume. Plant Physiol. **64**: 1083-1088

Pedalino, M., K.E. Giller, and J. Kipe-Nolt, 1992. Genetics of physiological characterization of non-nodulating mutant of *Phaseolus vulgaris* L. - NOD125. J. Exp. Bot. **43 (251)**: 843-849

Pedalino, M. and J. Kipe-Nolt, 1993. Common bean (*Phaseolus vulgaris* L.) mutants defective in root nodule formation. I. Physiological characterization. J. Exp. Bot. **44**: 1007-1014

Pedalino, M., J. Kipe-Nolt, L. Frusciante, and L. Monti, 1993. Common bean (*Phaseolus vulgaris* L.) mutants defective in root nodule formation. II. Genetic analysis. J. Exp. Bot. **44**: 1015-1020

Pemberton, I.J., G.R. Smith, and J.C. Miller, 1990. Inheritance of ineffective nodulation in cowpea. Crop Sci. **30**: 568-571

Penmesta, R.V. and D.R. Cook, 1997. A legume ethylene-insensitive mutant hyperinfected by its rhizobial symbiont. Science. **275**: 527-530

Peoples, M.B., D.F. Herridge, and J.K. Ladha, 1994. Biological nitrogen fixation: an efficient source of nitrogen for sustainable agricultural production? Plant and Soil. **174** : 3-28

Pierce, M. and W.D. Bauer, 1983. A rapid regulatory response governing nodulation in soybean. Plant Physiol. **73**: 286-290

Postma, J.G., E. Jacobsen, and W.J. Feenstra, 1988. Three pea mutants with an altered nodulation studied by genetic analysis and grafting. J. Plant Physiol. **132**: 424-430

Pracht, J.E., C.D. Nickell, and J.E. Harper, 1993. Genes controlling nodulation in soybean:Rj_5 and Rj_6. Crop Sci. **33**: 711-713

Pracht, J.E., C.D. Nickell, J.E. Harper, and D.G. Bullock, 1994. Agronomic evaluation of non-nodulating and hypernodulating mutants of soybean. Crop Sci. **34**: 738-740

Purdom, D. and A.T. Trese, 1995. Morphological and molecular characteristics of host-conditioned inefficient root nodules in cowpea. Plant Physiol. **109**: 239-244

Rolfe, B.G. and P.M. Gresshoff, 1988. Genetic analysis of legume nodulation. Annu. Rev. Plant Physiol. Plant Mol. Biol. **39**: 297-319

Rupela, O.P., 1992. Natural occurence and salient characters of nonnodulating chickpea plants. Crop Sci. **32**: 349-352

Sagan, M. and G. Duc, 1996. *Sym28* and *Sym29*, two new genes involved in regulation of nodulation in pea (*Pisum sativum* L.). Symbiosis (Rehovot). **20**: 229-245

Sagan, M. and P.M. Gresshoff, 1996. Developmental mapping of nodulation events in pea (*Pisum sativum* L.) using supernodulating plant genotypes and bacterial variability reveals both plant and *Rhizobium* control of nodulation regulation. Plant Sci. **117**: 167-179

Sagan, M., T. Huguet, D. Barker, and G. Duc, 1993a. Characterization of two classes of non-fixing mutants of pea plants (*Pisum sativum* L.). Plant Sci. **95**: 55-66

Sagan, M., A. Messager, and G. Duc, 1993b. Plant symbiotic mutants as a tool to analyse nitrogen nutrition and yield relationship in field-grown peas (*Pisum sativum* L.). Plant and Soil. **153**: 33-45

Sagan, M., T. Huguet, and G. Duc, 1994. Phenotype characterization and classification of nodulation mutants of pea (*Pisum sativum* L.). Plant Sci. **100**: 59-70

Schauser, L., A. Roussis, J. Stiller, and J. Stougaard, 1999. A plant regulator controlling development of symbiotic root nodules. Nature. **402**: 191-194.

Schmidt, J.S., J.E. Harper, T.K. Hoffman, and A.F. Bent, 1999. Regulation of soybean nodulation independent of ethylene signaling. Plant Physiol. **119**: 951-959

Schulz, S., J.D.H. Keating, and G.J. Wells, 1999. Productivity and residual effects of legumes in rice-based cropping systems in warm-temperate environment. I. Legume biomass production and N fixation. Field. Crops Res. **61**: 23-35

Schulz, S., J.D.H. Keating, and G.J. Wells, 1999. Productivity and residual effects of legumes in rice-based cropping systems in warm-temperate environment. II. Residual effect on rice. Field. Crops Res. **61**: 37-49

Shantharam, S. and A.K. Mattoo, 1997. Enhancing biological nitrogen fixation: an appraisal of current and alternative technologies for N input into plants. Plant and Soil. **194**: 205-216

Sheehy, J.E., 1999. Relationship between senescence, photosynthesis, nitrogen fixation and seed filling in soybean *Glycine max* (L.) Merr. Ann. Bot. **51**: 679-682

Sherrier, D.J., A.Y. Borisov, I.A. Tikhonovich, and N.J. Brewin, 1997. Immunocytological evidence for abnormal symbiosome development in nodules of the pea mutant line Spring2Fix - (*sym31*). Protoplasma. **199**: 57-68

Shritliffe, S.J. and J.K. Vessey, 1996. A nodulation (Nod+/Fix-) mutant of *Phaseolus vulgaris* L. has nodule-like structures lacking peripheral vascular bundles (*Pvb-)* and is resistant to mycorrhizal infection (Myc-). Plant Sci. **118**: 209-220

Shritliffe, S.J., J.K. Vessey, B.R. Buttery, and S.J. Park, 1996. Comparison of growth and N accumulation of common bean (*Phaseolus vulgaris* L.) cv. OAC Rico and its two nodulation mutants, R69 and R99. Can. J. Plant Sci. **76**: 73-83

Sidorova, K.K., N.Y. Gordienko, and T.I. Novikova, 1995a. Ultrastructure of the root nodules in symbiotic mutants of pea, *Pisum sativum*. Tsitologiia. **37**: 849-852

Sidorova, K.K., V.K. Shumny, and L.P. Uzhintseva, 1995b. Genetic experiments with pea mutants pending symbiosis studies. In: Proc. 10th Intern.Cong. on Nitrogen Fixation. Kluwer Academic Publisher, Dordrecht. pp.475-478

Sidorova, K.K. and V.K. Shumny, 1998. Analysis of pea (*Pisum sativum* L.) supernodulating mutants. Russian J. Genet. **34**: 1233-1235

Sidorova, K.K., V.K. Shumny, and E.Y. Valsova, 1997. Study of pea symbiotic mutants. Genetika. **33**: 656-659

Sidorova, K.K. and L.P. Uzhintseva, 1992. Use of mutants to identify genes controlling symbiotic characteristics in pea. Genetika. **28**: 144-151

Sinclair, T.R. and C.T. de Wit, 1975. Photosynthesis and nitrogen requirement for seed production by various crops. Science. **189**: 565-567

Singh, I. and M.H. Ahmad, 1991. Competitive interaction between non-nodulating and nodulating strains for nodulation of cowpea (*Vigna unguiculata*). FEMS Microbiol. Letters. **81**: 157-160

Singh, O. and O.P. Rupela, 1998. A new gene that controls root nodulation in chickpea. Crop Sci. **38**: 360-362

Singh, O., H.A. van Rheenen, and O.P. Rupela, 1992. Inheritance of new nonnodulation gene in chickpea. Crop Sci. **32**: 41-43

Song, L., B.J. Carroll, P.M. Gresshoff, and D.F. Herridge, 1995. Field assessement of supernodulating genotypes of soybean for yield, N_2 fixation and benefit to subsequent crops. Soil Biol. Biochem. **27**: 563-569

Sprent, J.I. and P. Sprent. (1990) Nitrogen Fixing Organisms: Pure and Applied Aspects. Chapman and Hall, London.

Streeter, J.G., 1988. Inhibition of legume nodule formation and nitrogen fixation by nitrate. CRC Crit. Rev. Plant Sci. **7**: 1-23

Suganuma, N., F. Kawate, and H. Tanaka, 1990. Adsorption of *Bradyrhizobium japonicum* and root hair curling in non-nodulating (rj_1 rj_1) soybean plant. Soil Sci. and Plant Nutr. **36**: 121-127

Suganuma, N. and T.A. LaRue, 1993. Comparisons of enzymes involved in carbon and nitrogen metabolism in normal nodules induced by a pea mutant E135 (*sym13*). Plant and Cell Physiol. **34**: 761-765

Suganuma, N., M. Tamaoki, and H. Kouchi, 1995. Expression of nodulin genes in plant-determined ineffective nodules of pea. Plant Mol. Biol. **28**: 1027-1038

Takahashi, M., M. Kokobun, and S. Akao, 1995. Characterization of nitrogen assimilation in a supernodulating soybean mutant En6500. Soil Sci. and Plant Nutr. **41**: 567-575

Toomsan, B., J.F. McDonagh, V. Limpinuntana, and K.E. Giller, 1995. Nitrogen fixation by groundnut and soybean and residual nitrogen benefits to rice in farmers' fields in Northeast Thailand. Plant and Soil. **175**: 45-56

Tsyganov, V.E., E.V. Morzhina, S.Y. Stefanov, A.Y. Borisov, V.K. Lebsky, and I.A. Tikhonovich, 1998. The pea (*Pisum sativum* L.) genes *sym33* and *sym40* control infection thread formation of root nodule function. Mol. Gen. Genet. **259**: 491-503

Vance, C.P., 1983. Rhizobium infection and nodulation: A beneficial plant disease? Annu. Rev. Microbiol. **37**: 399-424

Vance, C.P., 1997. Enhanced agricultural sustainability through biological nitrogen fixation. In: Biological Fixation of Nitrogen for Ecological and Sustainable Agriculture. Legocki, A., H. Bothe and A. Pühler (Eds.) Springer Verlag, Berlin. pp.179-186

Vest, G., 1970. Rj_3: a gene controlling ineffective nodulation in soybean. Crop Sci. **10**: 34-35

Vest, G. and B.E. Caldwell, 1972. Rj_4: a gene controlling ineffective nodulation in soybean. Crop Sci. **10**: 34-35

Vincent, J.M., 1980. Factors controlling the legume-*Rhizobium* symbiosis. In: Nitrogen Fixation. Vol. 2. W.E. Newton and W.H. Orme-Johnson (Eds.) Univ. Park Press, Baltimore. pp. 103-129

Vuong, T.D., C.D. Nickell, and J.E. Harper, 1996. Genetic and allelism analyses of hypernodulation soybean mutants from two genetic backgrounds. Crop Sci. **36**: 1153-1158

Wani, S.P., O.P. Rupela, and K.K. Lee, 1995. Sustainable agriculture in the semi-arid tropics through biological nitrogen fixation in grain legumes. Plant and Soil. **174**: 29-49

Williams, L.F. and D.L. Lynch, 1954. Inheritance of non-nodulating character in soybean. Agron. J. **46**: 28-29

Wu, S. and J.E. Harper, 1991. Dinitrogen fixation potential and yield of hypernodulating soybean mutants: a field evaluation. Crop Sci. **31**: 1233-1240.

Summary

Introduction

The major problem of sustainable agriculture is to overcome the range of stresses in environmentally compatible ways. To do this, genetic variability is needed in the best genotypes that are improved for the areas of need. Thus, induced mutation technologies and biotechnologies are major players in creating the desired variability for stress resistance or tolerance and in manipulating and evaluating them in efficient ways for improved germplasm. The problems of biotechnological applications in developing countries are many-fold and the approaches can be basically divided into *in vitro* culture applications and molecular approaches. There are both common and individual problems associated with the application of induced mutation technologies.

Mutation techniques are well established and there are a number of them to choose from for application to a particular species and situation. There is no doubt that the production and use of mutants will increase on a world-wide scale. There is a strong and growing demand by breeders, to produce new and useful variation in many species. Additionally, the scientific community has become intensely interested in mutations as they offer ideal tools for basic research in understanding gene function. Mutants have direct application in the emerging disciplines of molecular physiology and functional genomics.

The transfer of technology to developing countries needs more than just products of biotechnology but the technology itself. There are several reasons for this. First, the best-adapted germplasm to be improved is usually the local germplasm, with specific traits acceptable to the consumers, and thus can best be selected by the local breeders. Secondly, it provides local employment and enhances local research capability, enabling them to also work on minor crops that otherwise would be ignored. However, this requires strong local government involvement and support. Often such support may not be feasible for a government and the development of regional centres of excellence may be desired.

Creation of variability, mutation techniques and functional genomics

Creation of variation is a major concern of geneticists and plant breeders and is closely tied to germplasm preservation and enhancement. Today major efforts are being made to identify and sequence all the genes in plants. To do this, mutations and allelic variation are necessary.

The sequencing of total genomic DNA and the partial sequencing of cDNAs derived from reverse transcription of transcribed mRNA (referred to as expressed sequence tags - ESTs) is providing information on the numbers and structure of plant genes expressed in different tissues and during different periods of the life cycle. Current estimates from such studies of the model species *Arabidopsis* indicate that plant species contain at least 21,000 unique genes. However, most of the sequences obtained cannot yet be related to a particular function and phenotype. The great challenge in biology is to establish the function of these genes so that they can be manipulated for crop improvement. This involves the establishment and validation of the function through genetic analysis, which is referred to as 'Functional Genomics'. An essential component of this system of validation will be the availability of defined mutants, and hence, mutation technologies are assuming an increasing importance in plant genomics. Thus, there is a need to further develop and characterize mutant phenotypes in agriculturally and horticulturally important species to aid this process. Activity is required in the following research areas:

- Mutational analysis of function and structure
- Defined deletion mutants for particular species
- Database of available mutant resources
- Allelic variation
- Population development

Mutational analysis of function and structure

In many species, research activity has aimed to develop a range of mutant phenotypes for desirable characteristics, for example; mutants for reduced height, stress tolerance and disease resistance. From these activities, in different laboratories, a great number of mutants have been identified within any particular species for a particular trait. However, there is a dearth of information on the allelism of different mutants and the numbers of independent loci that have been mutated. This information is required if candidate genes are to be mapped to particular loci and validated as the gene sequence, which is mediating the variation. To do this in a systematic way for a particular species and character, for example, in wheat, rice and barley, will require new mutagenesis programs for the development of mutants in a standard genetic background, preferably one which has been identified as being 'mutable'. This will require coordinated efforts between different laboratories. Eventually, the availability of such mutants will enable ESTs for candidate genes to be mapped onto panels of characterized mutants for particular loci, which will enable the relationships between gene structure, function and allelic variation to be established.

Defined deletion mutants for particular species

The mapping of candidate gene cDNA probes usually relies on identifying RFLP in a segregating population. However, this requires that the probe reveals RFLP in the particular population(s) that is available. Frequently this is not the case and the intra-chromosomal location of candidates can be difficult to establish. However, if the sequence is absent, as in an aneuploid or a deletion line, then a presence/absence polymorphism will be exhibited and RFLP is not required. Thus, if a systematic set of deletion lines covering the whole genome is available in a particular species, then all candidate gene sequences could be physically mapped in the genome, and their function associated to the change in phenotype apparent in the deletion mutant line. This will enable the provisional characterization of function of all sequences available, which can be further verified by complementation analysis of the mutant by creating a transgenic line for the candidate gene in the mutant line. Indeed, if overlapping deletions are available, fine physical mapping will be possible. Generally, such an analysis will work well in a polyploid species such as wheat, which can tolerate aneuploidy, but it may be difficult to obtain a complete deletion set for a diploid such as rice and barley.

Thus, there is a need to create comprehensive sets of deletion lines in a range of model and agriculturally important species using appropriate mutagens. This can be achieved either via radiation or chemical mutagenesis. A strategy that can be implemented to do this is to generate a large M_1 population and then to self-individual M_1 plants to create M_2 progenies. A sample of M_2 progeny per M_1 plant (say 20) are then advanced by single seed descent to M_5/M_6 or doubled haploidy to create a population of neo-homozygous, or completely homogygous mutant deletion lines, respectively. These can then be characterized by using a set of defined molecular markers which have comprehensive marker coverage over the whole genome, and thus will identify a complete set of lines which is systematically nullisomic for the whole genome. Once such a set is established it can be used to partition the genome into a set of defined 'deletion bins' into which all candidate DNA sequences (cDNAs, ESTs, small insert libraries, large insert libraries) can be mapped. Indeed, with developments in DNA chip technology it should be possible to create 'mutant chips' which can be used for the automated physical mapping of candidate gene sequences.

Another method of characterizing genes sequences by function is to create mutants by insertional mutagenesis, either by transposon tagging or T-DNA transformation, and hence directly relating phenotype to gene sequence. Evidence suggests that low-level irradiation treatments can activate retro-transposon activity in some species, and there is a need to develop methodologies to investigate whether activation of such elements is possible.

Database of available mutant resources

Mutation research over this last century has produced an enormous number of mutant phenotypes in a very wide range of species. However, much of the information available on these mutants - pleiotropic effects, gene locations, allelism tests - is contained in laboratory note books and is not widely available to the scientific community. There is a great need to establish comprehensive electronic databases of mutants so that the information can be collected from a variety of sources and then effectively disseminated. These databases of mutants can then be included into existing crop databases, e.g. GrainGenes, RiceGenes.

Allelic variation

A continued supply of allelic variation is necessary for continued crop improvement in terms of agronomic performance, tolerance to biotic and abiotic stresses and value added traits. Numerous studies have shown that the gap between the primary gene pool of most crops and their secondary and tertiary gene pools is gradually widening. Efforts must continue to provide the variability from several aspects.

Screening of accessions in all gene pools is an ongoing activity to provide additional alleles not only for conventional cytogenetic and breeding approaches but additional alleles for the genomics approaches. Recent screening activity in wheat has been for variability for traits that are complex in inheritance such as resistance to tanspot, eyespot, take-all root rot, BYDV, WSMV, Fusarium head blight and abiotic stresses. For introgressions from tertiary gene pools it is essential that translocated segments are small to minimize linkage drag and that they are compensating, i.e. involve homoeologous chromosomes. This can now be monitored by *in situ* hybridization and/or markers. Induced mutations should probably receive more attention in terms of generation of additional variability that they can now be manipulated more precisely, e.g. utilizing site-directed mutagenesis and precisely characterizing mutants down to the nucleotide level using new techniques.

Population development

Populations for molecular mapping and gene tagging purposes can be constructed using haploid methodology or single seed descent (SSD). The latter works well when controlled environment conditions are available to grow four or more generations per year. Haploid methodology based either on chromosome elimination techniques or anther or microspore culture is available in most crops and accessible by developing as well as developed countries.

Molecular tools for research support

There are many types of molecular tools available which are suitable for distinct purposes. This section will elaborate on such tools, their applications and limitations where appropriate.

Map construction

Microsatellites or SSRs

These are tandem repeats of 1, 2, 3, 4, 5, or 6 nucleotides that are amplified by PCR using unique flanking sequences as primers. The most useful SSRs contain at least 20 nucleotides in total. They can be produced by screening published plant DNA sequences or by screening libraries that are enriched for sequence repeats and then sequencing the clones. The primer sequences can be designed using computer programs such as PRIMER and then synthesizing the primer oligos. In other words there is considerable developmental work involved in detecting, synthesizing and mapping these loci. They may go across species such as hexaploid wheat to tetraploid wheat but not as frequently across genera. Thus, they should be synthesized for each genus or species and therefore are not too amenable to comparative mapping. Microsatellites are highly abundant in plant genomes (estimates in rice genome are 5,700-10,000 microsatellites). They are uniformly distributed over the genomes in rice, corn, wheat and barley but clustered near the centromeres in tomato.

The basis of polymorphism is variation in repeat length, *i.e.* the number of units in a repeat. Usually an average of two alleles are detected per microsatellite but up to 25 alleles have been detected per locus in rice and barley. They are inherited in a codominant manner. The multiplex ratio (number of markers that can be generated in a single reaction) is 1 but they are amenable to multiloading so that 5-6 primers can be loaded simultaneously into a single reaction tube and 5-6 loci screened simultaneously.

Microsatellite maps have been produced in wheat, barley, rice and corn. In wheat up to 1000 markers are available and up to 400 in barley. Large numbers have also been produced in maize but mainly are proprioratory. The applications of microsatellites include studies of DNA fingerprinting and variety protection, germplasm conservation and utilization, gene and QTL analysis, pedigree analysis and marker assisted breeding and screening of large insert libraries.

Considering the various attributes of microsatellites such as abundance, random distribution, ability to detect numerous alleles, known map locations, codominant inheritance and multiple loading possibilities, they have to be considered as a preferable technique for marker assisted selection studies. There is a nominal fee for the license to use these markers for commercial applications such as variety development.

In cases such as wheat, barley, rice where microsatellite sequences have been published it is a matter of synthesizing the oligonucleotides. This can be done commercially for a cost of approximately US $50.00 per sample. This sample however is sufficient for 1,000 reactions so the sample can be shared with other laboratories or with developing countries. As pointed out above if sequence data is not available it must be acquired from a sequence database or by screening libraries enriched for sequence repeats. This is a costly and labourious exercise. In terms of the reaction, DNA extraction is relatively simple and similar to the RAPD reaction. 50

to 100 ng of DNA is required compared to 20 ng for RAPD. A vertical denaturing gel system is required. Initial screening can be done on agarose gels but to detect differences of several nucleotides, thick polyacrylamide gels are required. Fluorescent detection methods are available which adds to the complexity. Generally speaking, a laboratory equipped to do simple RAPDs can be upgraded to analyse microsatellites.

AFLP

AFLP relies on selective amplification of small genomic restriction fragments produced by double digestion with a rare-cutting restriction endonuclease (e.g. *Eco*RI or *Pst*I) plus a frequent cutter such as *Mse*I onto which known sequence adapters have been ligated. Digested and ligated DNA is preamplified using an *Eco*RI directed primer. Preamplification primers can contain selective nucleotides at 3' ends. Preamplified products are then used as templates for selective amplification and additional selective nucleotides can be added (up to 3) to 3' ends.

Polymorphisms in AFLP systems are based on base changes such as insertions and deletions. The multiplex ratio is very high, so that 50-100 markers can be generated in a single reaction. Expression of bands are usually dominant.

For marker assisted selection (MAS), individual polymorphic bands must be excised and cloned since it becomes too expensive to analyse every individual in a population with a full set of primers. (These are purchased as kits from KeyGene and a licence fee is required for use of materials for commercial purposes). Banding patterns can be detected within either ^{33}P or silver staining. Using the latter detection method, it is easier to isolate a band from a gel for purposes of cloning and sequencing.

This technique can be used for: identifying markers tightly linked to quantitative genes and QTLs. In one study QTLs identified by AFLP occurred in the same, or adjacent to, regions detected by other marker systems. Linkage maps can be constructed with AFLP, e.g., as in Avena, Arabidopsis, Petunia, Populus. In this case, since chromosomal locations of markers are not known such maps may need to be anchored by RFLPs, for example. AFLPs can be useful for filling gaps in existing maps. Other uses include fingerprinting plant and bacterial genomes, examining gene pool variation, fine scale mapping of mutations, aligning YACs and BACs into contigs for chromosome walking and producing tight linkages to desired genes for the chromosome landing approach.

AFLP analysis is based on kits and instructions supplied by representatives of KeyGenes such as Life Technologies. Kits cost $1,000 and will perform 60 reactions. Three kits would be sufficient to map 600 loci on a population of 200 lines. DNA extraction is similar to that for the RAPD method except that 0.5 to 1 ug of DNA is required per reaction. Long vertical polyacrylamide gels in a denaturing system are required. Double digestion is required followed by preamplification then selective amplification. Detection is done by either silver staining or use of ^{33}P. Overall this is a fairly complex technique but can be mastered in fairly short order by technicians who are familiar with RAPDs and RFLPs.

RAPD

This method employs oligos with random sequences as primers. It generates dominant markers and detects medium levels of polymorphism. Polymorphism is based on insertions and deletions. RAPD markers are easier to convert to SCARS than are RFLP. There are problems with RAPDs in obtaining repeatable results between genotypes and laboratories.

Other PCR methods

There are other less known PCR methods that are variations of the above two methods. Inter-Repeat Amplification (IRA) used 3' or 5' anchored microsatellite primers to amplify subsets of inter repeat regions. It gives a high multiplex ratio similar to AFLP. Another technique involves restricting the amplified product to reveal polymorphisms that may not have existed with the amplicon (CAPS - cleaned amplified polymorphic sequence). The Selective Amplification of Polymorphic Loci (SAMPL) uses one AFLP primer and one self-anchoring mircosatellite primer to amplify random genomic sequences that contain SSRs. This method combines the advantages of AFLPs and microsatellites.

RFLP

This is a reliable, accurate method of detecting polymorphisms. Numerous linkage maps have been constructed by this method and since these markers work well across genotypes it has permitted the comparative mapping that has been done to date. It reveals moderate levels of polymorphism compared to AFLPs. Despite being expensive, it is still probably the most accurate method for construction of maps and gene tagging. For marker assisted selection studies, RFLP markers need to be converted to PCR based markers (STS) but this has proven to be more difficult than anticipated.

Other methods

Single nucleotide polymorphism is a developing, revolutionary technology that will detect single base changes. Retrotransposon sequences combined with microsatellite primer systems are an emerging technology for fingerprinting in particular.

Use of molecular maps

Maps, if constructed with common markers such as RFLP, will facilitate comparative mapping not only between species but also genera. Comparative mapping enhances studies of species evaluation, gene tagging (qualitative & QTL) and supply of markers between species. The existence of extensive molecular maps in species such as *Secale cereale*, *Hordeum vulgare*, *T. tauschii*, *T. monococcum*, *Ae. umbellulata* will add a great deal of precision to the integration of traits from those species into wheat. Where in the past only major genes with obvious phenotypes had been integrated, now QTLs from alien species can be integrated and their epistatic interactions measured. Maps constructed from PCR based systems are still the best means of obtaining information on species that have not been genetically analysed. Map data and information on tight linkages plus sequence information can be used to compare mutants to parents. Extensive databases are being assembled based on comparative mapping in the Triticeae. The GrainGenes database at the Cornell University contains marker data on wheat, barley, oats, maize, rice, sugarcane and soybean. The EGRAM database, which is now being assembled in Europe will combine linkage maps of wheat, barley, oat, ryegrass and maize. Microcloning of single chromosomes or chromosome arms can be used to produce targeted saturated libraries.

EST (Expressed Sequence Tag) databases

These are beginning to be assembled as components of genomics initiatives. The rice genome program now has more than 30,000 ESTs and in other species, collectively, hundreds of thousands are available. Most are privately owned but accessible by negotiation.

BAC-YAC libraries

There are small insert libraries that can be aligned using ESTs, AFLP etc., and are used for chromosome walking for gene isolation. They are amenable to comparative mapping and gene cloning. Scientists in Japan have constructed YAC contigs in one-half of the rice genome for purposes of gene isolation and YAC or BAC libraries are now available in barley, wheat, rice, maize and *T. monococcum*.

Analogue technology

Genes have zones of common sequences, e.g., leucine-rich repeats and nucleotide binding sites that can be used to synthesize primers and amplify analogues from other species. Structures of cloned genes from model systems such as *Arabidopsis* can be used to construct primers and amplify analogues for other species for a variety of traits such as value-added parameters, flowering time, dwarfing, biotic stresses, etc. Such sequences can be used to study gene clusters and to isolate and clone genes.

Transformation technology

Dozens of gene constructs are currently available for transformation in cereals, e.g., fungal/pest resistance genes such as antifungal proteins, protease inhibitors, peroxidase genes, Bt genes antisense constructs, virus resistance genes such as coat protein genes, antiviral proteins, and value-added traits such as wheat glutenin genes, pea lipoxygenase and starch branching enzymes. Most of these genes are constrained by intellectual property. Similarly, most of the promoters used to drive these genes are privately owned. However, additional promoters occasionally, arise from the public domain, e.g., such as the tCup promoter that was recently announced.

Methods for gene transfer

Conventional methods for gene transfer from interspecific or intergeneric hybrids involves hormone applications and embryo rescue. Direction recombination occurs between homologous genomes. Between homoeologous genomes recombination may be induced by suppressing the *Ph* genes, using callus cultures or gamma radiation. In the latter system additional manipulations may be required to integrate the smaller segment of chromatin to minimize linkage drag.

FISH-GISH technology

Fluorescent *in situ* hybridization (FISH) or genomic *in situ* hybridization (GISH) use either a small DNA fragment or total genomic DNA, respectively, labelled with fluorescent probe(s) so that genes or genomes within a species can be differentiated or intergenomic translocations characterized. By similar means, BACs and YACs can be physically mapped. Unique sequences of up to 1 Kb have been mapped in transgenic plants and are important time savers in determining the number of insertion sites and in selecting plants that are homozygous for the transgenes. Gaps in YAC and BAC contigs can be quantified by fibre-FISH on extended chromatin fibres. These labelling tools are now important molecular adjuncts in genome mapping, for characterization of germplasm and in cytogenetic analyses. More than one colour of fluorescent probe can be used at one time, i.e. multicoloured or McFISH.

Requirements for marker assisted selection and future directions

A quick and dirty DNA extraction method or seed biopsy method (non-destructive) is required for DNA extraction, mutant characterization and marker-assisted selection. PCR methods such as microsatellites are amenable to multiplexing so that 10-12 genes or QTLs can be screened and pyramided simultaneously. Continued public effort is essential to enlarge gene pools by mutant induction, tissue culture technology and introgression through wide crosses.

Fingerprinting cultivars should continue and be expanded by targeting gene-rich regions, QTL rich regions, and hot spots for recombination in preparation for monitoring gene introgression. Continued, shared efforts are essential to develop and deploy new marker systems, e.g. shared library screening and mapping followed by sharing of synthesized primers for applications. Merging of maps should be undertaken to locate as many alleles as possible for particular traits.

In vitro methodology

The development of tissue culture systems with high regeneration capacity has been a limitation to the application of many molecular and biotechnology tools in crops, particularly the cereals. Being able to regenerate plants from single cells is a critical step for protoplast fusion, production of haploids from gametic cells, genetic transformation and in the creation of somaclonal variation. The advantages and applications of these tools will be considered.

Haploid methods

Haploid production procedures have two main advantages, namely; the saving of time in various breeding and research procedures and in producing genetically homozygous lines. They are valuable in many areas such as:

- Rapid development of pure line cultivars or inbreds for hybrids. It permits the fixation of transgressive segregants from hybrids exhibiting heterotic effects.
- Speeding up recurrent parent selection when backcrossing valuable mutant traits or alleles into good genotypes and for recurrent selection in breeding.
- Developing populations to be used for mutant trait analysis (QTL) and for molecular marker, molecular mapping and molecular breeding.
- As large uniform haploid single-cell targets for plant transformation.
- For mutation and selection in plants, either as direct microspore targets or in producing homozygous mutants (DH lines) from M_1 plants by various DH procedures.

There are a number of procedures for producing haploid plants, which fall into female gamete sources (wide hybridization/chromosome elimination and ovule culture from female gametes) and anther or isolated microspore culture using male gametes. Available suitable procedures vary with crop species. The most productive haploid procedure is isolated microspore culture in some Brassica species and in barley (and recently wheat). Wide hybridization with *Hordeum bulbosum* followed by selective chromosome elimination is a well-established system for haploid production in barley. Similarly in hexaploid wheat, maize pollen provides reasonable production of haploids across genotypes but needs to be perfected on tetraploid wheat. Chromosome doubling is also necessary when using the chromosome elimination systems. In *japonica* rice, anther culture is relatively easy and efficient while good procedures are lacking for *indica* genotypes. Haploidy techniques for maize and most legume crops are essentially lacking while ovule culture works in sugar beets.

There is widespread use of DH techniques for cultivar production in Brassica, tobacco, barley, rice and wheat. It is used so widely that numbers of cultivars are impossible to determine. In barley, Devaux et al. (1996) counted at least 60 cultivars from DH techniques. DH production is also feasible in other crops such as sunflower, potatoes and asparagus.

Future efforts on DH Technology should:

- Concentrate on further improvement of procedures to obtain more efficient and repeatable DH production for crops where these are lacking and that can be transferred to locations of need.
- Consider the establishment of good facilities for donor plant growth or, where not feasible, establish central facilities where DH lines could be produced by the best procedures for a given crop.
- Provide training programs in established labs. DH procedures require excellent training and persistent practice to achieve repeatable results. Training programs are the most effective way to transfer the technology, followed by short subsequent refresher or updating periods.

Doubled haploid procedures will continue to evolve and to become valuable and more widespread as tools in plant breeding, mutation and crop research. They will be important for the needs of increased quality and production of crop foods and for rapid incorporation of new environmentally friendly germplasm.

Protoplast fusion and transformation

Protoplast fusion has unique applications that are not readily achieved by other means. It is most often used to produce hybrids between species that are not sexually compatible, usually distantly related. The object is to transfer genes, most often for stress or pest resistance, into crops that are lacking such resistance in their germplasm. Such fusions are often symmetrical and the cytoplasm of the crop species is chosen. However asymmetrical fusions are also of interest in transferring a nucleus from one species into the cytoplasm of another species or genotype. Most examples of asymmetric fusion deal with the transfer of cytoplasmic male sterility. Good examples of such research are described in more detail in the paper by Kinoshita et al. found in this volume. The procedures are intricate and labourious.

Protoplasts have also been used for direct DNA uptake for gene transformation in various crop species. By removal of the cell wall, it is possible to use systems to disrupt the plasma membrane, allowing the DNA to enter the protoplast and be taken up into the host genome. Electroporation apparatus are commercially available for such work and chemicals such as polyethylene glycol (PEG) will also allow DNA to enter through the membrane. While successful, this method of gene transformation is not often preferred when compared to other procedures.

Somaclonal variation

When various plant tissues or cells are placed into tissue culture, they are often stimulated to undergo cell division by the addition of plant hormones in combination with a sugar energy source and other critical nutrients. This growth is found most often in a disorganized morphological structure called callus. Most often callus can be induced to regenerate from meristematic areas if not left too long (more than 6 months) in culture. This passage through a callus culture system usually generates variations in the phenotype and genotype compared to the original genotype and is thus a source of induced variation. Some of the variation may be transient and others are shown to be permanent and transmitted to subsequent generations.

The cause and nature of this genetic variation has been the subject of much speculation and debate. Nevertheless, it provides a good source of variation, particularly for plants that reproduce asexually (i.e. do not normally reproduce through the meiotic sexual cycle). The plants produced from somaclonal variation can then be subjected to the stress for which the variation is being sought, either under field conditions or in the tissue culture media under which the callus is being produced.

Genetic transformation

Under the ideal situation, is it desirable to put the gene of interest into a single cell, which is capable of regenerating into a plant, so that the whole plant will carry the gene of interest as opposed to only a portion of the cells, i.e. a chimeric plant. Thus, various types of cell cultures are used as targets for plant transformation. Very little is known about how the gene is incorporated into the host cell genome and thus the approaches are most often quite random as is the site at which the gene is incorporated. However, approaches are being developed to incorporate the gene into a specific site in the host genome based upon recombination with homologous sequences in the host genome.

Nevertheless, the main problem is a low frequency of transformation due to the low frequency of cells in the target that receive the gene and the subsequent problem of regenerating the plant from these cells. Utilizing modifications of the system developed in nature for inserting genes into plant cells, *Agrobacterium tumifaciens*, it has been possible to infect plant cells with *Agrobacterium* and have it transfer the gene of interest into the plant cells. This has been well developed for various legume crops that are readily infected by the *Agrobacterium*, but only recently has this system worked on monocot species. Thus, the biolistic method of shooting the DNA into cells on metal particles has been the main procedure for monocot transformation. Such a crude system often damages the cells and thus, the regeneration of transgenic plants from the culture is a major concern.

Application of technologies in developing countries

Mutation techniques

Mutation techniques are well established and there are a number of techniques to choose from for application to a particular species and situation. There is no doubt that the production and use of mutants will increase on a world-wide scale. There is a strong and growing demand by breeders, in many species, as a means of producing new and useful variation. Additionally, the scientific community has become intensely interested in mutations as they offer ideal tools for basic research in understanding gene function. Mutants have direct application in the emerging disciplines of molecular physiology and functional genomics. The deployment of mutants raises serious questions on mutant germplasm production, identification, characterization, collection, maintenance, distribution and the world-wide accessibility of the information.

It is strongly recommended that much of this should be breeder driven in the regions of need. Breeders are encouraged to:

- Expand the development of new mutants in locally adapted germplasm
- Characterize the mutants
- Deposit the new mutant germplasm into accessible collections
- Provide information for accessible data bases
- Collaborate with functional genomics research centres, and other interested scientists working mainly on physiological aspects of stress responses.

Mutant and germplasm collections and databases have already been established for many crops including barley, rice, wheat, and tomato. This should be expanded to include new mutant lines from collaborators. Consideration should be given to the establishment of mutant germplasm collections and databases for other regionally important crops around the world. There are many crops, where no central collection site exists; the initiation of such collections should be encouraged. It is understood that collection of all extant mutants is not feasible (there are too many), but a start can be made by maintaining the most important ones. Additional funding should be a high priority to support collections. Screening mutations for responses to stress has been and continues to be an important area on a world-wide basis. A great deal more work is needed to improve crop plants for sustainable agriculture in regions where stress tolerance is critical for improved production in order to meet increasing needs. This work cannot stop and must be advanced. A major aim is the development of quick, cheap, simple, and reliable tests, which correlate with field performance. This would supply breeders with efficient, rapid techniques that would speed up variety production. Again, this requires a major input by plant breeders, agronomists, and laboratory based scientists – there have been far too many instances where physiological theories on stress tolerance have proved inappropriate to field performance. It should be noted that mutants, that control plant responses to one stress, may have positive effects on other stresses; these mutants should therefore be exposed to wider testing. Markers for mutants are essential, it is envisaged that more DNA markers will be developed, but simple morphological markers are also adequate.

Plant Biotechnology

The problems of biotechnological applications in developing countries are many-fold. The technology can be basically divided into *in vitro* culture applications and molecular approaches. There are common problems associated with application of these two groups and also specific problems.

Molecular markers

Among the seed propagated crops, the biotechnological applications in cereals and some legumes (soybean, pea and beans) with regard to gene mapping and molecular marker techniques are being well advanced. Nevertheless, it is important that mapping populations are developed in locally adapted material. This is critical in assessing the effects of mutated genes on traits important to the breeder and location. Genetic maps and markers are required for this but do not involve the development of new markers, because these can be transferred from established maps/genomic centres. Gene mapping and molecular marker based techniques in many tropical legumes and oil seeds are in the initial phases of development and therefore more research is needed before their practical application.

Tissue culture applications

Much more research is desired in tissue culture applications of many tropical cereals, legumes and oil seeds. This includes, for example, doubled haploid technology of *indica* rice, maize, millets, legumes, etc.

Transfer of technology

Transfer of technology to developing countries and the role of governments and multinational seed companies was discussed. During the transfer of technology, the governments should negotiate with companies on equal terms, so that not only the products of biotechnology are transferred, but also the technology itself. For example, a condition may be laid, that employment of nationals of the developing country is obligatory. This will enhance the local research capability. This applies to major crops where seed companies have interest. As far as the minor crops are concerned, government support is important. As one government may not be able to support such projects, setting up of regional centres of excellence is desired. Many governments of developing countries are reluctant to fund biotechnology research because they are not convinced of cost-effectiveness. The possibility of funding by producers association was discussed and it was agreed that this is possible in export crops such as tea, rubber, banana or coffee, but not in subsistence crops. For competent negotiation, the governments need the advice of trained legal personnel on patenting and other related issues. This knowledge is also lacking in many developing countries.

The patenting of many useful genes, technologies etc. by private companies was viewed as a major obstacle in transfer of technology to developing countries. Here, the only solution in the foreseeable future would be improving the negotiating power of these countries by training legal manpower. There exist international agencies that can help developing countries in such negotiations. International institutes can also play an important role.

Transgenic crops and the developing countries

Lack of control over entry of transgenic crop cultivars and food products into the third world countries is another problem. The respective governments should therefore start and speed up setting up of legal procedures of control over transgenic plants, transgenic products and technologies. This is all the more important in view of the fact that the escape of transgenes in developing countries, where most of our major food crops have originated, can be disastrous, as they harbour large numbers of related wild species. It is already evident that marketing of transgenic products by developing countries to the developed countries will also face problems. Therefore, before embarking on the development of transgenic products, it is advisable to understand the market trends, consumer acceptability and government regulations of importing countries.

Changing training needs

Many developing countries have already developed their own expertise in mutation techniques. However, there exist training needs for associated novel techniques such as doubled haploids, DNA marker methods and marker-assisted techniques. There is currently a major effort by several multinational biotechnological companies to exploit the utilization of molecular breeding for improved agricultural production. These efforts are targeted at a very small number of commercially important crops and will in all probability have a major impact only in the developed countries. The concern of the group was that the technology gap between developed and developing countries will become larger. International and national efforts should be strengthened to encourage and speed up technology transfer into regions in developing countries where the new emerging technologies will have considerable impact on sustainable agriculture systems.

Index